Future Trends in Ultra-Precision Machining

Future Trends in Ultra-Precision Machining

Guest Editors

Changlin Liu
Yanbin Zhang
Xiaoliang Liang

MDPI

Basel • Beijing • Wuhan • Barcelona • Belgrade • Novi Sad • Cluj • Manchester

Guest Editors

Changlin Liu
School of Mechanical Science and Engineering
Huazhong University of Science and Technology
Wuhan
China

Yanbin Zhang
School of Mechanical and Automotive Engineering
Qingdao University of Technology
Qingdao
China

Xiaoliang Liang
School of Mechanical Engineering
Shandong University
Jinan
China

Editorial Office
MDPI AG
Grosspeteranlage 5
4052 Basel, Switzerland

This is a reprint of the Special Issue, published open access by the journal *Micromachines* (ISSN 2072-666X), freely accessible at: https://www.mdpi.com/journal/micromachines/special_issues/YW185K02EA.

For citation purposes, cite each article independently as indicated on the article page online and as indicated below:

Lastname, A.A.; Lastname, B.B. Article Title. *Journal Name* **Year**, *Volume Number*, Page Range.

ISBN 978-3-7258-7681-5 (Hbk)
ISBN 978-3-7258-7682-2 (PDF)
https://doi.org/10.3390/books978-3-7258-7682-2

Contents

About the Editors

Changlin Liu

Changlin Liu is an Associate Researcher at the School of Mechanical Science & Engineering, Huazhong University of Science and Technology. He completed his PhD studies in Mechanical Engineering at Huazhong University of Science and Technology. After that, he worked as a postdoctoral fellow at State Key Laboratory of Ultra-precision Machining Technology in The Hong Kong Polytechnic University. His interests include ultra-precision machining of brittle materials, atomic numerical simulation and field-assisted machining. He has published more than 40 peer-reviewed research papers, and the total citations of these publications are more than 1000 in Google Scholar (h-index = 20).

Yanbin Zhang

Yanbin Zhang is a professor at Qingdao University of Technology, China. He received his Ph.D. degree from Qingdao University of Technology, China, in 2018. He obtained the project support of 'Hongkong Scholars' and worked in The Hong Kong Polytechnic University, China, from 2021 to 2023. He is a Highly Cited Researcher of Clarivate and a young expert of Taishan Scholars in Shandong Province of China. He is the Associate Editor of *Intelligent and Sustainable Manufacturing* and *Journal of Advanced Manufacturing Science and Technology*, Young Editorial Board Member of *International Journal of Extreme Manufacturing* (Q1, Top), *Friction* (Q1, Top) and *Journal of Mechanical Engineering* (EI). His current research interests focus on intelligent and clean precision manufacturing. He has published 153 papers, which are cited more than 14098 times in total. He won the Hiwin Doctoral Dissertation Award and Shandong Province Natural Science Award 2nd Prize.

Xiaoliang Liang

Xiaoliang Liang is a professor at the School of Mechanical Science and Engineering at Shandong University. He completed his PhD in Mechanical Engineering at Shandong University. He then engaged in postdoctoral research at the State Key Laboratory of Ultra-Precision Machining Technology at the Hong Kong Polytechnic University. His interests include high-quality and efficient cutting machining, process quality prediction, surface/interface friction physical and chemical properties and modification control. He has published more than 60 peer-reviewed research papers, which have a total of more than 2000 citations in Google Scholar (h-index = 23).

micromachines

MDPI

Editorial

Editorial for Future Trends in Ultra-Precision Machining

Changlin Liu [1,*], Yanbin Zhang [2,*] and Xiaoliang Liang [3,*]

1 School of Mechanical Science and Engineering, Huazhong University of Science and Technology, Wuhan 430074, China
2 School of Mechanical and Automotive Engineering, Qingdao University of Technology, Qingdao 266520, China
3 School of Mechanical Engineering, Shandong University, Jinan 250061, China
* Correspondence: charlie_lcl@outlook.com (C.L.); zhangyanbin1_qdlg@163.com (Y.Z.); liangxiaoliang@sdu.edu.cn (X.L.)

1. Introduction for This Special Issue of Future Trends in Ultra-Precision Machining

Ultra-precision machining (UPM) is an advanced manufacturing technology used to produce components with exceptionally high dimensional accuracy, surface integrity, and form precision [1,2]. In general, it refers to machining processes capable of achieving tolerances in the sub-micrometer range and surface roughness at the nanometer level, which forms the backbone and support of today's innovative technology industries from optoelectronics, aerospace, optics, and biomedical engineering [3,4].

With the rising demand for precision devices in harsh environments, achieving UPM on brittle materials has become of great interest in research of advanced manufacturing. During machining, surface defects and subsurface damage such as cracks, fractures, and tool wear caused by their brittle nature is the main reason behind the deterioration of the machined surface quality of brittle materials [5–7]. Scholars have revealed that defect-free surfaces on brittle materials can be fabricated via ductile machining by decreasing the material removal thickness to a nanoscale level [8], while the machining mechanism is not fully revealed since microstructures such as crystal orientation and grain boundaries have significant influence on deformation behavior.

To overcome the undesirable machining defects and improve the machineability of brittle materials, various field-assisted machining technologies have been proposed, which introduce controllable external energy fields (such as laser heating [9], ultrasonic tool vibration [10], magnetic fields [11], etc.) to regulate the local mechanical response of the material or the removal mode of the machining process, ultimately improving the machined surface quality and machining efficiency. Furthermore, to amplify the effects of applied assistive fields, researchers have coupled the energy fields and explored the mechanisms in multi-field-assisted cutting [12]. In field-assisted cutting, the introduced energy field alters both the intrinsic properties of materials and the tool–workpiece interaction. Consequently, the underlying mechanisms differ fundamentally from those of conventional processes and single-field-assisted machining, making this a prominent research hotspot in the current field of UPM.

In recent years, Artificial Intelligence (AI) has become a pivotal core technology to promote the intelligent transformation of ultra-precision machining systems, aiming to address challenges such as highly coupled process interactions, nanometer-level precision requirements, limited process observability, and the high cost caused by trial-and-error experiments [13–15]. It is worth noting that digital twin technology enables a closed-loop integration including state perception, process prediction, and decision optimization, thereby providing system-level understanding and control for complex machining processes. At the same time, the rapid advancement of foundation models, such as Large Language Models [16], Vision Foundation Models [17] and Industrial Time-series Foundation Models [18], has provided new technical approaches for manufacturing knowledge modeling, cross-process knowledge transfer, and human–machine collaborative decision-making. Compared to traditional methods that often rely on empirical parameter tuning, simplified analytical models, or process-specific modeling, these developments are expected to fundamentally enhance process predictability, controllability, and system-level intelligence in next-generation ultra-precision manufacturing systems.

Building on sincere cooperation with the Guest Editors, this Special Issue concentrates on research in the field of ultra-precision machining, including removal mechanisms, field-assisted machining, and AI-driven process optimization and measurement. As depicted in Figure 1, this Special Issue featured a diverse array of topics, publishing a total of 8 contributions, comprising 7 original research articles and 1 review paper, which is briefly introduced below according to three aspects (i.e., atomic mechanism, field-assisted machining, AI-driven measurement and process optimization) of UPM in this Special Issue.

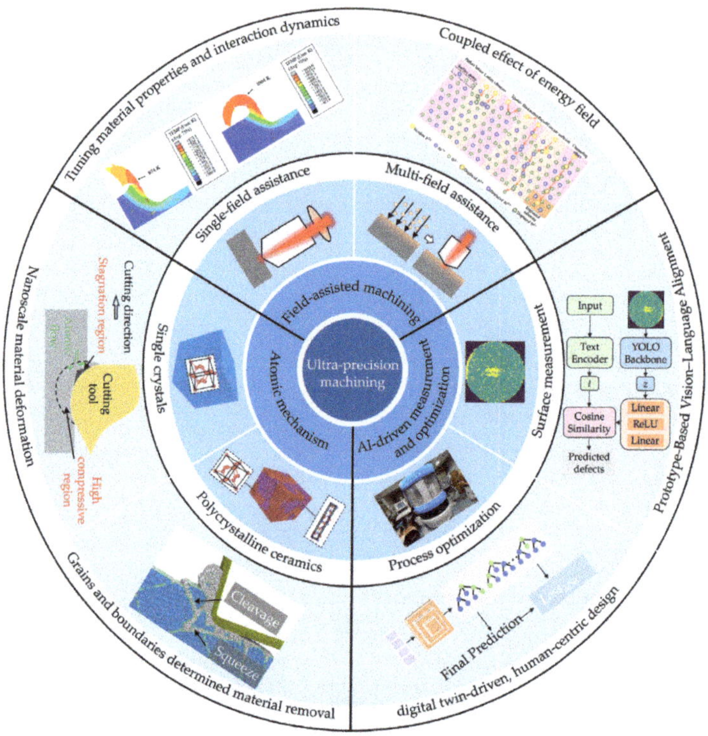

Figure 1. Topics covered in the Special Issue titled "Future trend in Ultra-precision machining".

2. Material Removal Mechanism of UPM in This Special Issue

The deformation mechanism of material in UPM differs from that in macroscale machining since the material removal thickness decreases into nanoscale, which is comparable to the cutting tool edge radius. Furthermore, microstructures such as grain boundaries and crystal orientation have significant influence on the material removal behavior. Due to the difficulty in direct observation of material deformation at the nanoscale level, molecular dy-

https://doi.org/10.3390/mi17040422

namics (MD) simulation [19,20], which is based on first-principles calculations or empirical interatomic potentials, has been widely used to describe the material deformation behavior in UPM owing to its advantages in simulating plastic deformation behavior such as shear flow [21], dislocation propagation [22], and phase transition [23]. The material removal mechanism of UPM covered in this Special Issue includes investigation and review of the deformation mechanisms of different brittle materials including single crystals, polycrystal ceramics, and amorphous glass.

For typical brittle materials, Sheng et al. [24] summarized previous research on the machining mechanism of brittle materials during single-point diamond turning and synthesized numerical and experimental findings across material classes, including single crystals, polycrystalline ceramics, and amorphous glass. They discussed the material removal mechanisms represented by ductile, transition, and brittle removal.

Differing from single crystals, the material removal mechanism in polycrystalline materials (such as reaction-bonded silicon carbide) is determined by grain distribution and boundaries. Mo et al. [25] conducted MD to investigate the effect of abrasive size on surface morphology and subsurface deformation mechanism of reaction-bonded silicon carbide during nano-grinding. They analyzed the surface swelling, material removal rate, and high-pressure phase transition under different grain sizes. These results provided the theoretical basis for an in-depth understanding in mechanisms during UPM of materials with random distribution of grains.

3. Field-Assisted Machining in This Special Issue

This Special Issue covers a wide range of investigations into single-field-assisted machining including laser-assisted cutting, and multi-field-assisted machining (e.g., laser-vibration assisted cutting, implantation-laser assisted cutting).

For brittle materials such as glass-ceramic optical components, the high-hardness and low-fracture toughness make it prone to cracks and subsurface damage during conventional cutting. Li et al. [26] investigated the dynamic removal mechanisms of glass–ceramics under laser-assisted nanoscale cutting conditions through numerical simulations and systematic experiments. They quantified the enhancement effect of laser power on the critical depth of no observed surface cracks (NOSC) and revealed the mechanisms through which laser assistance inhibits crack propagation. The findings provide theoretical support for optimizing laser-assisted cutting parameters and achieving high-quality machining of glass–ceramics.

For sapphire crystals, which have extraordinary hardness and are widely used in advanced optics, microelectronic devices, and medical instruments. The improvement in machineability by a single assistive field is limited. Ke et al. [27] used ion implantation and laser-assisted cutting to improve the machining performance of C-plane sapphire. Their groove cutting experiments verified the enhancement in ductile machinability of the modified sapphire under LADM. At a laser power of 16 W, the ductile–brittle transition depth of the modified sapphire increased to 450.67 nm, representing a 51.57% improvement over conventional cutting. The findings of this study provide valuable insights for improving the ductile machining performance of hard and brittle materials.

During multi-field-assisted machining, the machining mechanism is complex since multiple interactions are involved in the machining process. Chu et al. [28] investigated the machining mechanism of single-crystal silicon under the combination of laser heating and tool vibration using MD simulations. They discussed the effect of tool vibration trajectory determined by different tool edge radii is discussed under the condition of raising temperature. Their results help to improve the understanding of machining mechanics in multi-field-assisted machining.

https://doi.org/10.3390/mi17040422

4. AI-Driven Measurement and Process Optimization in Special Issue

This Special Issue focuses on Artificial Intelligence (AI)-based surface defect measurement [16] and process optimization of the ultra-precision machining, aiming to effectively control the production process, reduce variation in the manufacturing process, and improve the yield rate represent important competitive factors for wafer factories.

With the rapid development of semiconductor manufacturing technology, methods to wafer bin maps provide valuable information for engineers to quickly identify potential root causes through accurate pattern recognition. Wang et al. [16] proposed YOLO-LA, a lightweight prototype-based vision–language alignment framework that integrates a pretrained frozen YOLO backbone with a frozen text encoder to enhance wafer defect recognition. They introduced a learnable projection head to map visual features into a shared embedding space, enabling classification through cosine similarity. The proposed framework is lightweight and suitable for real-time industrial wafer inspection systems.

During UPM, maintaining the machining quality presents a greater challenge when working conditions change due to the nanoscale material removal and surface roughness requirement. Mo et al. [29] conceptually proposed a digital twin (DT)-driven, human-centric design framework that integrates key factors of multi-jet polishing process. They introduced a feature-encoded transfer learning-based model to enhance surface roughness prediction accuracy and robustness under varying working conditions. Their model offers a practical and extensible perspective for optimizing complex ultra-precision manufacturing processes under data-scarce and uncertainty-dominated conditions.

With the continuous development of semiconductor manufacturing technology and information technology, the sizes of wafer chips are becoming smaller and the variety is increasing, which has put forward high requirements for wafer chip precision manufacturing and packaging workshops. Wang et al. [30] researched the wafer chip precision packaging workshop rescheduling problem under events of machine breakdown, emergency order inserting, and original order modification. They proposed a mathematical model for the addressed problem and a hybrid algorithm combining an improved firefly optimization framework. Their model is effective and stable and is superior to the current advanced algorithms.

To conclude, we would like to acknowledge all the authors for their contributions to the success of this Special Issue in *Micromachines*, as well as the reviewers whose feedback helped to improve the quality of the published papers.

Conflicts of Interest: The authors declare no conflict of interest.

References

1. Fang, F.; Lai, M.; Wang, J.; Luo, X.; Yan, J.; Yan, Y. Nanometric cutting: Mechanisms, practices and future perspectives. *Int. J. Mach. Tools Manuf.* **2022**, *178*, 103905. [CrossRef]
2. Yip, W.S.; To, S.; Zhou, H. Current status, challenges and opportunities of sustainable ultra-precision manufacturing. *J. Intell. Manuf.* **2021**, *33*, 2193–2205. [CrossRef]
3. Zhang, X.; Hu, H.; Wang, X.; Luo, X.; Zhang, G.; Zhao, W.; Wang, X.; Liu, Z.; Xiong, L.; Qi, E.; et al. Challenges and strategies in high-accuracy manufacturing of the world's largest SiC aspheric mirror. *Light Sci. Appl.* **2022**, *11*, 310. [CrossRef] [PubMed]
4. Huang, L.; Wu, H.; Cai, G.; Wu, S.; Li, D.; Jiang, T.; Qiao, B.; Jiang, C.; Ren, F. Recent Progress in the Application of Ion Beam Technology in the Modification and Fabrication of Nanostructured Energy Materials. *ACS Nano* **2024**, *18*, 2578–2610. [CrossRef
5. Yan, J.; Asami, T.; Harada, H.; Kuriyagawa, T. Crystallographic effect on subsurface damage formation in silicon microcutting *CIRP Ann.-Manuf. Technol.* **2012**, *61*, 131–134. [CrossRef]
6. Yan, J.; Asami, T.; Harada, H.; Kuriyagawa, T. Fundamental investigation of subsurface damage in single crystalline silicon caused by diamond machining. *Precis. Eng.* **2009**, *33*, 378–386. [CrossRef]

https://doi.org/10.3390/mi17040422

7. Guo, J.; Zhang, J.; Pan, Y.; Kang, R.; Namba, Y.; Shore, P.; Yue, X.; Wang, B.; Guo, D. A critical review on the chemical wear and wear suppression of diamond tools in diamond cutting of ferrous metals. *Int. J. Extrem. Manuf.* **2020**, *2*, 012001. [CrossRef]
8. Peter, N.; Blake, R.S. Ductile-Regime Machining of Germanium and Silicon. *J. Am. Ceram. Soc.* **1990**, *73*, 949–957.
9. He, W.; Liu, C.; Xu, G.; Zhang, J.; Xiao, J.; Chen, X.; Xu, J. Effect of temperature on ductile-to-brittle transition in diamond cutting of silicon. *Int. J. Adv. Manuf. Technol.* **2021**, *116*, 3447–3462. [CrossRef]
10. Shamoto, E.; Moriwaki, T. Study on Elliptical Vibration Cutting. *CIRP Ann.* **1994**, *43*, 35–38. [CrossRef]
11. Yip, W.S.; To, S. Tool life enhancement in dry diamond turning of titanium alloys using an eddy current damping and a magnetic field for sustainable manufacturing. *J. Clean. Prod.* **2017**, *168*, 929–939. [CrossRef]
12. Zhang, J.; Li, X.; Ma, S.; Lu, Y.; Tian, H.; Xiao, J.; Xu, J. High efficiency fabrication of Si microlenses by applying in-situ laser and ultrasonic vibration hybrid diamond cutting. *Precis. Eng.* **2025**, *93*, 324–333. [CrossRef]
13. Wang, R.; Cheung, C.F.; Wang, C. Heterogeneous hypergraph learning for analyzing surface defects in additive manufacturing process. *J. Manuf. Syst.* **2024**, *76*, 1–10. [CrossRef]
14. Wang, R.; Cheung, C.F.; Wang, B.; Zhu, Z.; Tse, D.Y. Enhanced transformer method coupled with transfer learning for surface defect segmentation of myopia control spectacle lenses. *Opt. Express* **2025**, *33*, 13848–13863. [CrossRef] [PubMed]
15. Wang, R.; Cheng, M.N.; Loh, Y.M.; Wang, C.; Fai Cheung, C. Ensemble learning with a genetic algorithm for surface roughness prediction in multi-jet polishing. *Expert Syst. Appl.* **2022**, *207*, 118024. [CrossRef]
16. Wang, Z.; Yang, Y.; Chu, J.; Zang, Y.; She, Z.; Fang, W.; Wang, R. YOLO-LA: Prototype-Based Vision-Language Alignment for Silicon Wafer Defect Pattern Detection. *Micromachines* **2025**, *17*, 67. [CrossRef]
17. Wang, R.; Cheung, C.F.; Jiang, Y.; Xuan, D.; Poon, C.S. SAMSort: Vision foundation model for sorting construction and demolition waste. *Dev. Built Environ.* **2025**, *24*, 100811. [CrossRef]
18. Liang, Y.; Wen, H.; Nie, Y.; Jiang, Y.; Jin, M.; Song, D.; Pan, S.; Wen, Q. Foundation Models for Time Series Analysis: A Tutorial and Survey. In *Proceedings of the 30th ACM SIGKDD Conference on Knowledge Discovery and Data Mining*; Association for Computing Machinery: New York, NY, USA, 2024; pp. 6555–6565. [CrossRef]
19. Plimpton, S. Fast parallel algorithms for short-range molecular dynamics. *J. Comput. Phys.* **1995**, *117*, 1–19. [CrossRef]
20. Thompson, A.P.; Aktulga, H.M.; Berger, R.; Bolintineanu, D.S.; Brown, W.M.; Crozier, P.S.; in 't Veld, P.J.; Kohlmeyer, A.; Moore, S.G.; Nguyen, T.D.; et al. LAMMPS—A flexible simulation tool for particle-based materials modeling at the atomic, meso, and continuum scales. *Comput. Phys. Commun.* **2022**, *271*, 108171. [CrossRef]
21. Liu, C.; Chu, J.; Chen, X.; Xiao, J.; Xu, J. Molecular dynamics simulation on structure evolution of silica glass in nano-cutting at high temperature. *Mol. Simul.* **2020**, *46*, 957–965. [CrossRef]
22. Liu, C.; Zhuang, Z.; Chen, J.; Yip, W.S.; To, S. Atomic-scale investigation of Pt composition on deformation mechanism of AuPt alloy during nano-scratching process. *Surf. Interfaces* **2023**, *40*, 103126. [CrossRef]
23. Liu, C.; Chen, X.; Zhang, J.; Zhang, J.; Chu, J.; Xiao, J.; Xu, J. Molecular dynamic simulation of tool groove wear in nanoscale cutting of silicon. *AIP Adv.* **2020**, *10*, 015327. [CrossRef]
24. Sheng, X.; Zhu, Z.; Liu, C. A Review of the Machining Mechanisms in Field-Assisted Cutting of Brittle Materials. *Micromachines* **2026**, *17*, 361. [CrossRef] [PubMed]
25. Mo, H.; Chen, X.; Luo, C.; Cai, X. Atomic Insight into the Nano-Grinding Mechanism of Reaction-Bonded Silicon Carbide: Effect of Abrasive Size. *Micromachines* **2025**, *16*, 1049. [CrossRef]
26. Li, J.; Ji, F.; Xu, F. Surface Integrity of Glass-Ceramics by Laser-Assisted Diamond Cutting. *Micromachines* **2025**, *16*, 1054. [CrossRef]
27. Ke, J.; Mo, H.; Ling, K.; Chu, J.; Chen, X.; Xu, J. Enhancing the Machinability of Sapphire via Ion Implantation and Laser-Assisted Diamond Machining. *Micromachines* **2025**, *16*, 1165. [CrossRef]
28. Chu, J.; Yang, Y.; Zang, Y.; Ke, J.; Wang, Z.; Chen, C.; He, J.; Xu, A.; She, Z. Investigation into Laser-Vibration-Assisted Cutting of Single-Crystal Silicon by Molecular Dynamics Simulation. *Micromachines* **2025**, *16*, 1411. [CrossRef]
29. Mo, H.; Chen, X.; Guo, L.; Zhang, Z.; Chen, X.; Chu, J.; Wang, R. A Transferable Digital Twin-Driven Process Design Framework for High-Performance Multi-Jet Polishing. *Micromachines* **2026**, *17*, 226. [CrossRef]
30. Wang, Z.; Fang, W.; Yang, Y. An Effective Hybrid Rescheduling Method for Wafer Chip Precision Packaging Workshops in Complex Manufacturing Environments. *Micromachines* **2025**, *16*, 1403. [CrossRef]

https://doi.org/10.3390/mi17040422

micromachines

MDPI

Article

Atomic Insight into the Nano-Grinding Mechanism of Reaction-Bonded Silicon Carbide: Effect of Abrasive Size

Honglei Mo, Xie Chen, Cui Luo and Xiaojiang Cai *

Shanghai Aerospace Control Technology Institute, Shanghai 201109, China; mohonglei1987@sina.com (H.M.); spwanx@163.com (X.C.); luocui16@163.com (C.L.)
* Correspondence: cxj0927@163.com

Abstract: Reaction-bonded silicon carbide (RB-SiC) is a high-performance ceramic material known for its excellent mechanical, thermal, and chemical properties. It contains phases with different mechanical properties, which introduce complex machining mechanisms. In the present work, molecular dynamics (MD) simulation was conducted to investigate the effect of abrasive size on the nano-grinding mechanism of RB-SiC. The surface morphology and subsurface deformation mechanism were investigated. The simulation results suggest that when a small abrasive is used, the surface swelling of SiC is primarily generated by the bending and tearing of SiC at the interfaces. As the abrasive radius increases, the surface swelling is mainly formed by Si atoms, which is identified as elastic recovery. Meanwhile, the material removal rate gradually decreases, and the depth of plastic deformation is obviously increased. Stocking of Si is more apparent at the interface, and obvious sliding of SiC grains is observed, forming edge cracks at the margin of the workpiece. In the subsurface workpiece, the high-pressure phase transition (HPPT) of Si is promoted, and the squeeze of disordered Si is obvious with more dislocations formed when larger abrasive is used.

Keywords: molecular dynamics simulation; nano-grinding; reaction-bonded silicon carbide; nanoscale machining mechanism; abrasive size

1. Introduction

Reaction-bonded silicon carbide (RB-SiC) is a high-performance ceramic material known for its excellent mechanical, thermal, and chemical properties. It is widely used in the optical mirrors of micro-electromechanical systems, high-energy laser modules, and photoelectric detectors [1]. To meet the surface accuracy requirements in these applications, nanoscale machining methods, including nano-grinding and nano-polishing, have attracted great research interest in the field of ultra-precision machining (UPM) [2,3]. Different to macroscale machining, the material removal thickness in UPM ranges from tens to hundreds of nanometers, which introduces distinct deformation mechanism of workpiece material [4]. The microstructures in RB-SiC, such as grain/phase boundaries, have a significant influence on the material removal process. Therefore, an in-depth investigation of the deformation mechanism at nanoscale is essential to reveal the physical nature in UPM of RB-SiC.

Due to the difficulty faced in the experimental observation of the material removal process, numerical simulation is widely used in investigations of the machining mechanism in UPM. Owing to the rapid development of computational science, molecular dynamics (MD) simulation has emerged as an effective implement to investigate nanoscale deformation features, including phase transition [5], dislocation [6], and cracks [7]. It has been widely used to explore the deformation mechanism of various materials in UPM [8,9]. For instance, Wang et al. [10] studied the removal behavior of single-crystal Si during nanometric cutting via MD simulation. They revealed that contact-induced amorphization of Si dominates the extrusion process, while strain-induced amorphization is responsible for shear removal. Meng et al. [11] investigated the friction behavior of single-crystal 3C-SiC during nano-scratching using MD simulation. They observed structure transition from order to disorder phases, coupled with dislocations in the subsurface workpiece in nano-cutting. Sahoo et al. [12] employed MD simulation to investigate the scratch damage of silica glass against a rigid diamond indenter. They found that densification and shear flow play a major role in the formation of scratch damage. Bhamra et al. [13] used MD simulation with a reactive force field to investigate the wear of single-crystal diamond tips sliding on α-quartz surfaces covered with water molecules. They found that the diamond wear on α-quartz is caused by the formation of C-O interfacial bonds and C-C cleavage, coupled with the diffusion of carbon into the substrate and oxidation.

For polycrystalline materials, the deformation mechanism can be different from that of single crystals. Pradhan et al. [14] studied the wear-resistance behavior in polycrystalline Pt-Au under a hard grinding ball using MD simulation. They discussed the effect of adding atomic concentrations of Au at the grain boundaries (GBs) on the wear process. Their results indicate that the segregation of Au to the GBs of nanocrystalline Pt leads to an improvement in the slab's anti-wear behavior, suggesting that the underlying wear mechanism is adhesion-dominated. Goel et al. [15] conducted an MD simulation of nanometric cutting on polycrystal Si and revealed that GBs facilitate the amorphization and formation of irregular nanogrooves on the machined workpiece. Liu et al. [16] investigated the material removal behavior of polycrystal 3C-SiC during nano-scratching using MD simulation. They revealed that grain geometry and GBs in the polycrystalline workpiece could affect the internal stress state and suppress the phase transition process, which causes different subsurface damage patterns with single crystals. Meng et al. [17] built an MD model to study the friction mechanism during the nano-grinding of hot-pressed SiC, which is composed of 3C-, 4H-, and 6H-SiC grains. They revealed that the machinability of hot-pressed SiC is determined by the microstructure-related intergranular slip motion and grain fragmentation. Recently, Liu et al. [18] conducted grooving experiments and MD simulation to study the machining mechanisms of RB-SiC under various temperatures. Their results indicate the significance of property mismatch between phases in nanoscale machining. For brittle materials such as Si and SiC, the material removal mechanism is determined by the tool geometry parameters, including rake angle and edge radius [19,20]. However, in the above-mentioned research, the effect of tool geometry on the material removal behavior of RB-SiC was not mentioned, and the coupling deformation mechanism of Si and SiC phases under different tool shapes has not been fully revealed.

In the present work, MD simulation was conducted to explore the nano-grinding mechanism of RB-SiC. The effect of abrasive size on the deformation behavior at the phase boundaries was discussed in detail. A visual image of the material removal process and subsurface damage evolution during the nano-grinding process are established with consideration of the property mismatch between phases. This study contributes to improving our understanding in the machining mechanism of RB-SiC during UPM and provides insight into the coupling deformation behavior of different phases in nanoscale/macroscale machining.

2. Methodology

Figure 1 shows the adopted nano-grinding model, which contains a deformable workpiece and a rigid diamond abrasive. The polycrystal structure of the workpiece was constructed using the Voronoi algorithm from Atomsk [21,22]. The same amounts of Si and SiC grains were set in the workpiece, with random distributions. The average size of the SiC grains is 15.2 nm. Following the classic setup in the MD simulation of UPM [23], the workpiece atoms were classified into three groups: the fixed group, the thermostat group, and the Newtonian group. A periodic boundary condition was applied in the y direction, while a fixed boundary condition was used in the x and z directions. The analytical bond order potential (ABOP) [24] was used to describe the atomic interaction in the simulation systems. It is widely used in simulations of UPM on Si and SiC [25,26] since it provides a desirable description of the mechanical properties of diamond and diamond-like structures [27,28]. The energy function of ABOP is described as follows:

$$U = \sum_{i>j} f_C(r_{ij})[V_R(r_{ij}) - \frac{b_{ij} + b_{ji}}{2} V_A(r_{ij})] \tag{1}$$

$$V_R(r) = \frac{D_0}{S-1}\exp[-\beta\sqrt{2S}(r - r_0)] \tag{2}$$

$$V_A(r) = \frac{SD_0}{S-1}\exp[-\beta\sqrt{2/S}(r - r_0)] \tag{3}$$

$$f_C(r) = \begin{cases} 1 & r < R - D \\ \frac{1}{2} - \frac{1}{2}\sin\left(\frac{\pi}{2}\frac{r-R}{D}\right) & |R - r| \leq D \\ 0 & R + D < r \end{cases} \tag{4}$$

$$b_{ij} = (1 + \chi_{ij})^{-1/2} \tag{5}$$

$$\chi_{ij} = \sum_{k(\neq i,j)} f_C(r_{ik})\exp[2\mu(r_{ij} - r_{ik})]g(\theta_{ijk}) \tag{6}$$

$$g(\theta) = \gamma[1+\frac{c^2}{d^2} - \frac{c^2}{d^2 + (\cos\theta + h)^2}] \tag{7}$$

where D_0 and r_0 represent the dimer energy and the bond length, respectively; S and β are determined based on the Pauling plot slope and ground-state oscillation frequency of the dimer; R and D specify the position and the width of the cutoff region, respectively; and variables h, γ, c, d, and 2μ are adjusted to describe the three-body interactions. Details of these parameters are presented in Table 1. The Large-scale Atomic/Molecular Massively Parallel Simulator (LAMMPS) [29] was used to perform the simulation. And the Open Visualization Tool (OVITO) [30] was applied to visualize and analyze the results. Detailed information of the simulation is present in Table 2.

Table 1. Parameters of ABOP.

	Si	**C**
D_0 (eV)	3.24	6.00
r_0 (Å)	2.232	1.4276
S	1.842	2.167
β (Å^{-1})	1.4761	2.0099
R (Å)	2.82	2.00
D (Å)	0.14	0.15
γ	0.114354	0.11233
c	2.00494	181.910

Table 1. *Cont.*

	Si	C
d	0.81472	6.28433
h	0.259	0.5556
2μ (Å^{-1})	0.0	0.0

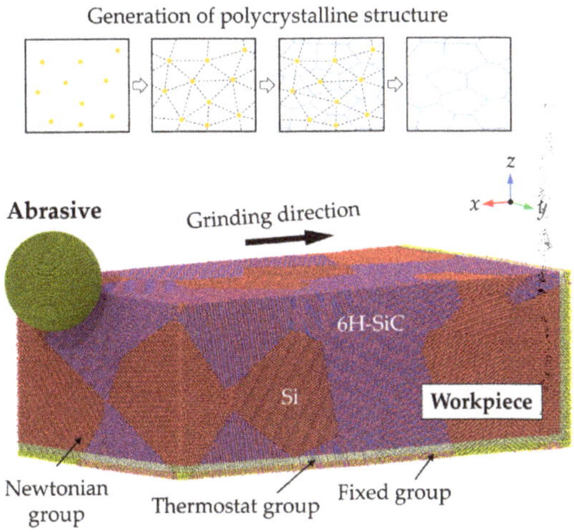

Figure 1. The adopted MD model for nano-grinding simulation.

Table 2. Details of the simulation parameters.

Parameters	Value
Workpiece dimension ($x \times y \times z$)	70 nm × 40 nm × 25 nm
Number of workpiece atoms	About 5 million
Abrasive radius (R)	5 nm, 6 nm, 7 nm, 8 nm, 9 nm, 10 nm
Grinding speed	50 m/s
Grinding depth	5 nm
Grinding distance	65 nm
Grinding temperature	300 K
Number of grains in workpiece	20 (10 Si grains and 10 SiC grains)

3. Results and Discussion

3.1. Surface Morphology

The height distribution and morphology of the machined workpiece are shown in Figure 2. The property mismatch between Si and SiC introduces obvious surface structures at the phase boundaries, which is distinct to the machined surface on single crystals [31]. Chips and side flow show an apparent irregular profile since different removal mechanisms are involved when machining Si and SiC phases. When small abrasives (R = 5 nm and 6 nm) are used, the surface swelling of SiC is observed as the abrasive moves from the SiC to the Si phase due to the bending and tearing of SiC at the interfaces. SiC nanocrystals are observed in chips because of the polycrystallization process. These hard nanocrystals are responsible for the formation of scratch patterns on the Si machined surface. As the abrasive radius increases, no obvious tearing of materials are observed at the interfaces, and the side flow is distributed more uniformly near the machined surface. Scratch patterns are still obvious, indicating the existence of SiC polycrystallization, while the surface swelling is mainly formed by Si atoms, which can be identified as the elastic recovery effect. It is revealed that for brittle materials such as Si and SiC, the highly compressed materials could be restored to their balanced position during unloading, causing deviation of the machined surface to the theoretical value [32].

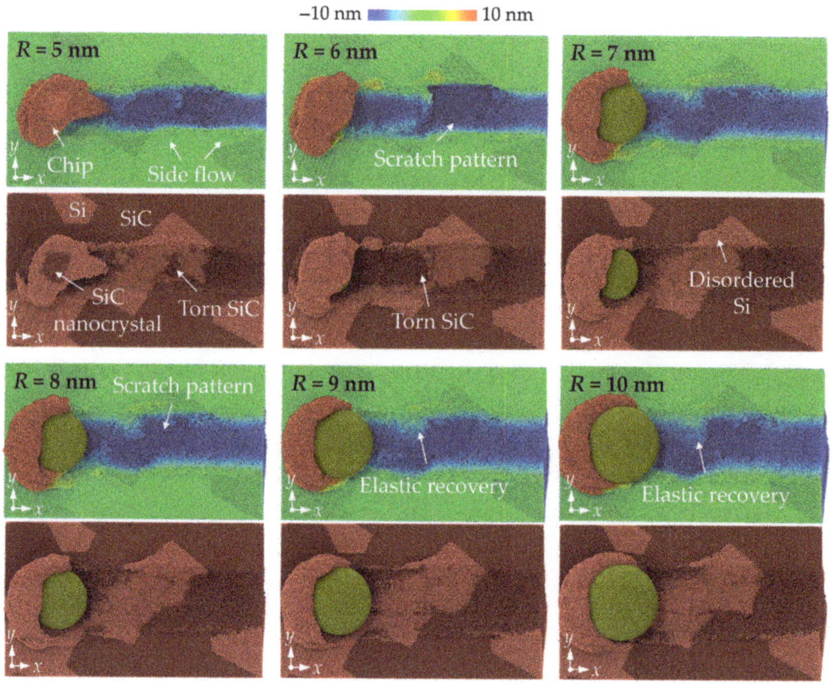

Figure 2. Height distribution and morphology of the ground surfaces machined by abrasives of different sizes.

The cross-sectional view of workpiece is shown in Figure 3, with the Si and SiC phases identified. In the MD simulation, the material removal thickness ranges from several to tens of nanometers, which indicates that the material is probably removed in the ductile mode. It can be concluded that the surface error is mainly caused by elastic recovery and scratch patterns due to the property mismatch between phases. When machining from SiC to Si, the torn SiC generated at the interface could move with the abrasive and collapse by the following SiC grain, while elastic recovery is more apparent when machining from the Si to SiC phase. When abrasives of a larger size are used, cracks are observed at the margin of the workpiece, which is caused by sliding of the SiC grains. To measure the effect of different swelling mechanisms on surface integrity, the surface roughness was calculated based on the positions of the surface atoms, which are identified according to the atomic volume (shown in Figure 4a). The surface atoms usually have a large atomic volume based on the Voronoi algorithm [33]. The surface roughness, *Sa*, can be estimated via the following equation [33]:

$$Sa = \frac{1}{A}\iint |z(x,y)|dxdy \approx \frac{1}{n}\sum |d| \quad (8)$$

where *d* represents the distance between surface atoms and the theoretical surface. Figure 4b shows the calculated roughness of the ground surface with abrasives of different sizes. The surface roughness shows a trend of increasing as the abrasive size increases, indicating the significance of elastic recovery on the overall surface roughness.

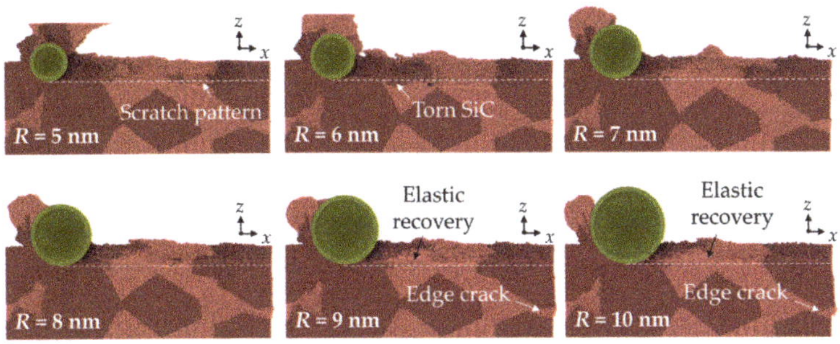

Figure 3. Cross-sectional view of the workpiece machined by abrasives of different sizes.

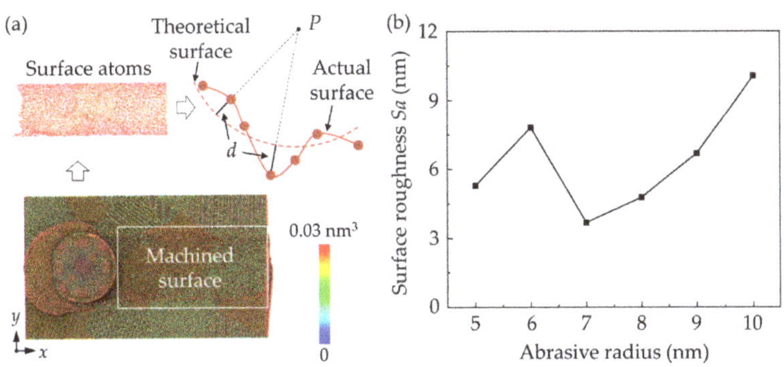

Figure 4. Surface roughness of the ground surface: (**a**) illustration of the surface roughness calculation; (**b**) the calculated roughness of the ground surface with abrasives of different sizes.

Figure 5 shows the calculated material removal rate as the grinding distance reaches 65 nm, which is calculated by counting the number of atoms in the deformation region. It is observed that compared to SiC, the Si phase has a higher removal rate due to lower hardness and brittleness. Meanwhile, the downward sliding of the SiC grains also contributes to the smaller material removal rate of SiC. As the abrasive size increases, the material removal rate gradually decreases. This variation can be explained by the fact that the sharp edges from smaller abrasives suppress the elastic deformation of the workpiece material and promote the concentrated plastic deformation in the machining region.

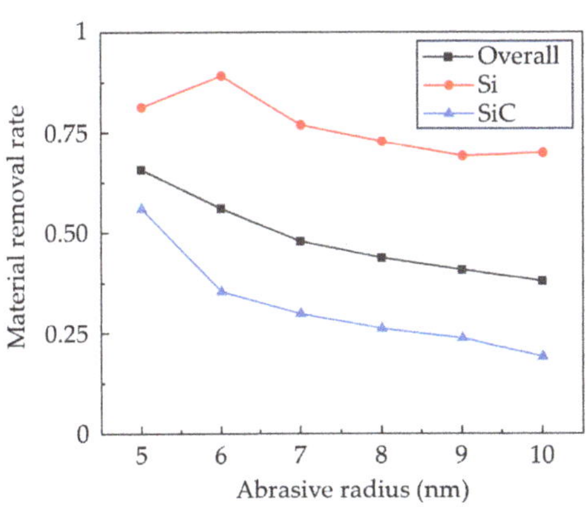

Figure 5. The material removal rate of the workpiece machined by abrasives of different sizes.

3.2. Deformation Behavior

During nano-grinding, the deformation of RB-SiC is more complex than that of single crystals due to the random distribution of grains and the property mismatch between phases. Figure 6 shows the cross-sectional views of the magnitude distribution of atomic displacement in workpieces, referring to the initial stage. Elastic deformation can be identified as the region where the displacement magnitude grows gradually, while plastic deformation is recognized as the clear interruption of the displacement magnitude since destruction of the crystal lattice occurs. When the abrasive cuts into the grains, the deformation mechanism is similar to that of single crystals. While as the abrasive cuts through the phase boundaries, the residual plastic deformation pattern on the machined surface is non-uniform due to the difference in deformation behavior between phases. It is observed that when machining from Si to SiC, the high-flowability disordered Si atoms are covered on the Si surface and stocked at the interfaces, while due to the insufficient support of Si, obvious bending and plastic deformation of the hard SiC occur when machining from SiC to Si. As the abrasive size increases, the depth of the plastic deformation is obviously increased. The stocking of Si is more apparent at the interfaces, which indicates stronger generation of the disordered phases. For the SiC phase, more dislocation patterns are observed inside the grains as a result of enhanced plastic deformation, and obvious grain sliding is observed, forming edge cracks at the margin of the workpiece.

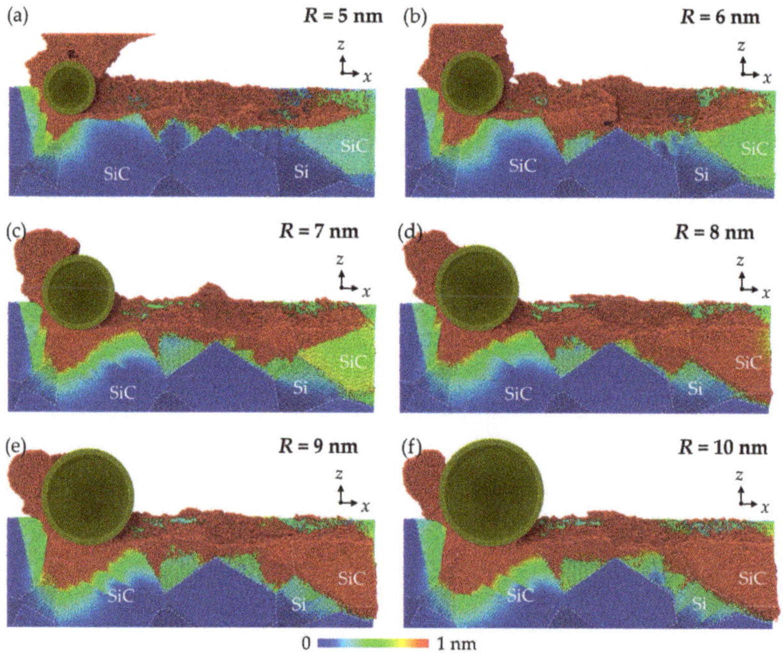

Figure 6. Cross-sectional views of atomic displacement magnitude distribution in workpieces machined by abrasives with radii of (**a**) 5 nm, (**b**) 6 nm, (**c**) 7 nm, (**d**) 8 nm, (**e**) 9 nm, and (**f**) 10 nm.

For a clear view of the crystal structure in the subsurface workpiece, the workpiece structure was identified via improved common neighbor analysis (CNA) [34], as shown in Figure 7. The white atoms are the disordered atoms, which mainly contain amorphous phase and defective structures. It can be seen that SiC nanocrystals are formed on the machined surface and chips due to the polycrystallization process, while most disordered phases are Si atoms generated by the high-pressure phase transition (HPPT). As the abrasive advances, the disordered Si atoms can stock and squeeze into the boundaries at the interfaces. When abrasives with larger size are used, the HPPT of Si is promoted, and the squeeze of the disordered Si becomes more obvious. Figure 8 shows the number of disordered atoms after grinding. It is observed that as the abrasive radius increases, more disordered Si atoms are generated, while the variation of disordered SiC atoms is slight. This result indicates that raising the abrasive radius enhances the HPPT of Si, while its influence on SiC is inapparent.

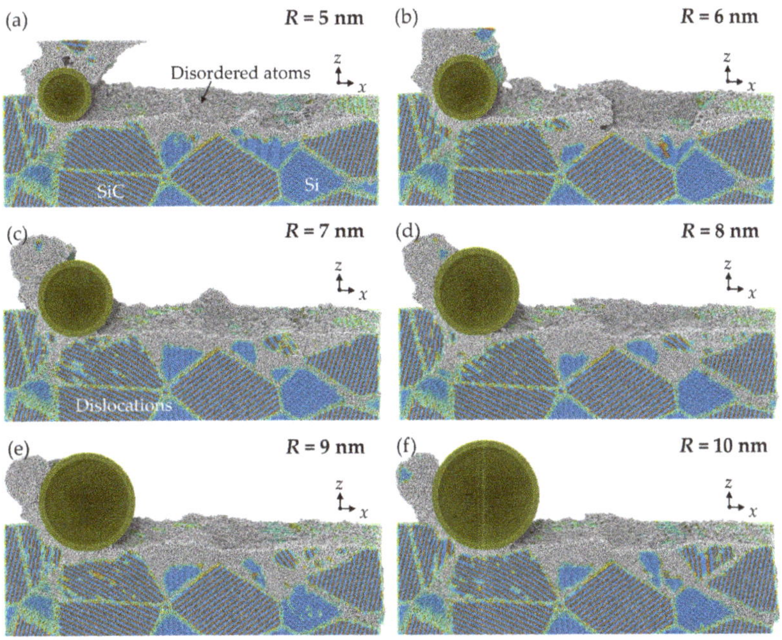

Figure 7. Cross-sectional view of the crystal structure in workpieces machined by abrasives with radii of (**a**) 5 nm, (**b**) 6 nm, (**c**) 7 nm, (**d**) 8 nm, (**e**) 9 nm, and (**f**) 10 nm.

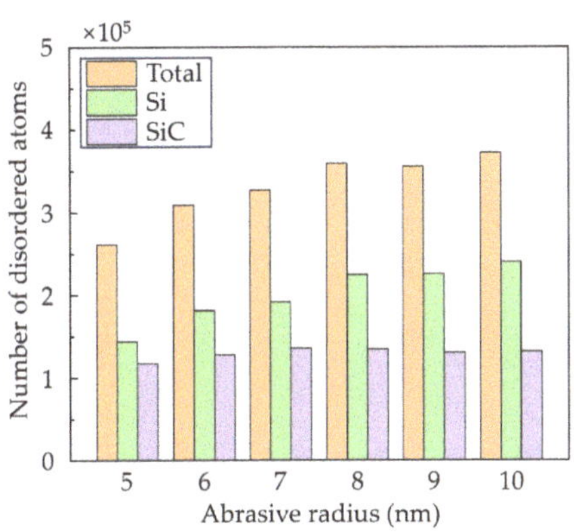

Figure 8. Number of disordered atoms in a workpiece machined by abrasives of different sizes.

3.3. Dislocation Analysis

From Figure 7, obvious dislocation patterns are observed in the subsurface workpiece. To give a clear view of the dislocation propagation during the nano-grinding of RB-SiC, the dislocation structures of Si and SiC atoms are calculated individually, as shown in Figure 9.

Atoms near the grain boundaries are deleted before calculation to eliminate the influence of the initial dislocation on the results. In Si grains, the dominant dislocation styles are perfect dislocation with a Burger's vector of 1/2<110>, and partial dislocation with a Burger's vector of 1/6<112>. It can be seen in Figure 9a that several dislocations are formed in the deformed region and the deep subsurface workpiece, which can be attributed to the compression from the deformed SiC grains; while for SiC grains, dislocations with Burger's vectors of 1/3<1-210> and 1/3<1-100> are observed in the subsurface workpiece as a result of polycrsytallization. The length of the dislocation is counted and is shown in Figure 10. It is clear that more dislocations are formed in the subsurface workpiece as the abrasive size increases, which is attributed to the larger deformed region induced by raising the abrasive size.

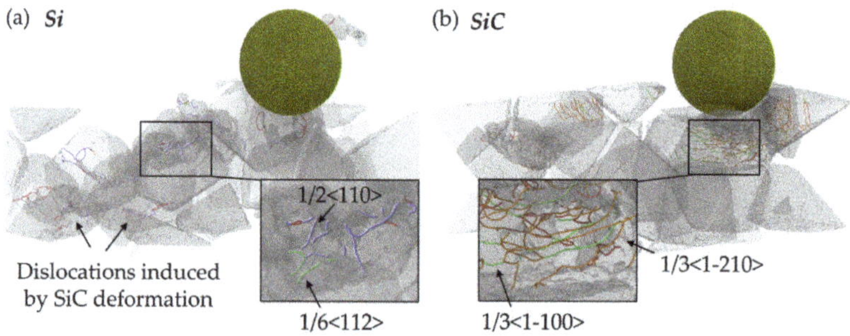

Figure 9. Snapshots of the dislocations in the (**a**) Si and (**b**) SiC grains of the workpiece machined by an abrasive with a 10 nm radius.

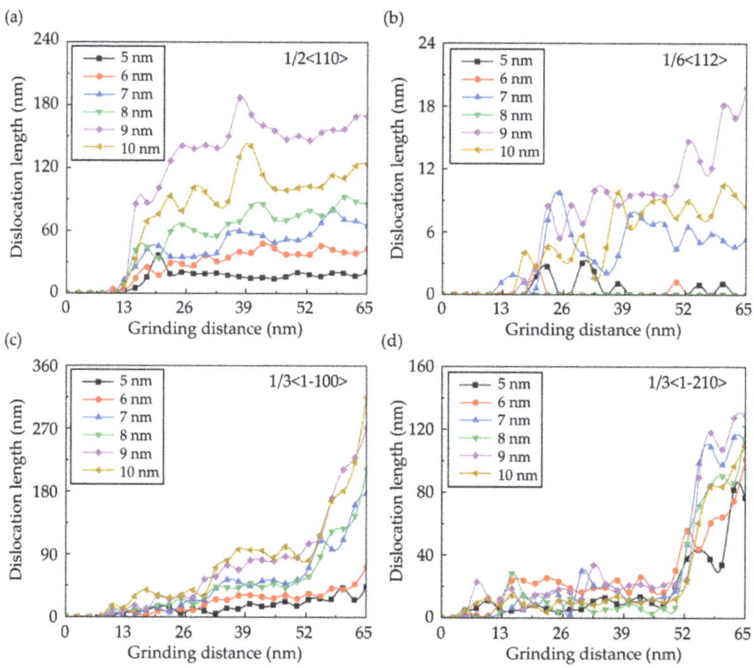

Figure 10. Variation of the dislocation length in a workpiece machined by abrasives with different radii: (**a**) 1/2<110> dislocations in the Si phase; (**b**) 1/6<112> dislocations in the Si phase; (**c**) 1/3<1-100> dislocations in the SiC phase; and (**d**) 1/3<1-210> dislocations in the SiC phase.

3.4. Internal Stress

Internal stress is intimately connected to the structural evolution in the subsurface workpiece during UPM. To further investigate the impact of abrasive size on subsurface damage during nano-grinding, the hydrostatic stress and von Mises stress of workpiece atoms are calculated using the following stress tensors [35]:

$$\sigma_{hydrostatic} = \frac{1}{3}(\sigma_{xx} + \sigma_{yy} + \sigma_{zz}) \tag{9}$$

$$\sigma_{vonMises} = \sqrt{\frac{1}{2}((\sigma_{xx} - \sigma_{yy})^2 + (\sigma_{yy} - \sigma_{zz})^2 + (\sigma_{zz} - \sigma_{xx})^2 + 6(\tau_{xy}^2 + \tau_{yz}^2 + \tau_{zx}^2))} \tag{10}$$

where σ_{xx}, σ_{yy}, σ_{zz}, τ_{xy}, τ_{xz}, and τ_{yz} are stress tensors. In this simulation, a cubic box spatial average of 2 nm is used to average the stress tensor. Figure 11 shows the stress distribution when machining across different phases. It can be seen in Figure 11a that the compressive stress increases obviously when the abrasive contacts with the hard SiC grains. Meanwhile, the bending of SiC grains causes an obvious increase in tensile stress inside the SiC grains and compressive stress at the interfaces. The increase in von Mises stress also indicates the bending of SiC grains. When the abrasive machines from the Si to the SiC phase (shown in Figure 11b), the compressive region is mainly restrained in the Si phase due to compression from the SiC and diamond abrasive. The slight increase in von Mises stress in SiC is caused by the enhanced extrusion of the disordered Si atoms. When the abrasive size increases, the compression of the workpiece material is significantly enhanced, causing more HPPT. When machining from SiC to Si phases, the bending of SiC becomes more apparent as an obvious increase in von Mises stress is observed in the SiC grains, as shown in Figure 11c. When machining from Si to SiC, the apparent increase in von Mises stress in SiC suggests that the extrusion of disordered Si atoms is enhanced (shown in Figure 11d), which introduces stronger upward atomic flow.

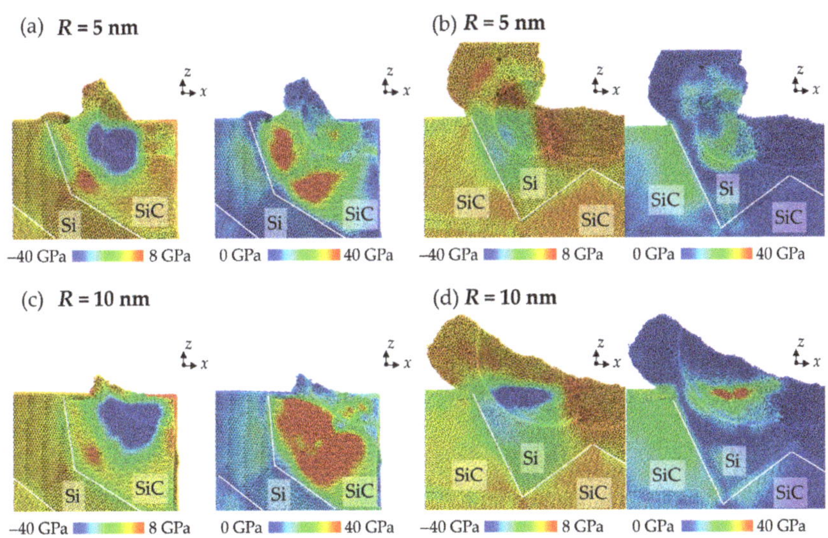

Figure 11. The stress distribution when machining across the phase boundaries with abrasives with radii of (**a**,**b**) R = 5 nm and (**c**,**d**) R = 10 nm.

To investigate the dynamic stress state, the average stress in workpiece during grinding was calculated, since varying the abrasive radius could change the scale of the deformation region. A slice with 1 nm thickness in the middle of the workpiece was selected to calculate the average stress, as shown in Figure 12. The calculated average stress is shown in Figure 13. The internal stress increases obviously when the abrasive contacts with the hard SiC grains. When the abrasive size increases, the internal stress raises obviously, especially when machining SiC grains. As the abrasive size increases from 5 nm to 10 nm, the average compressive stress increases by 184%, and the average von Mises stress increases by 140%.

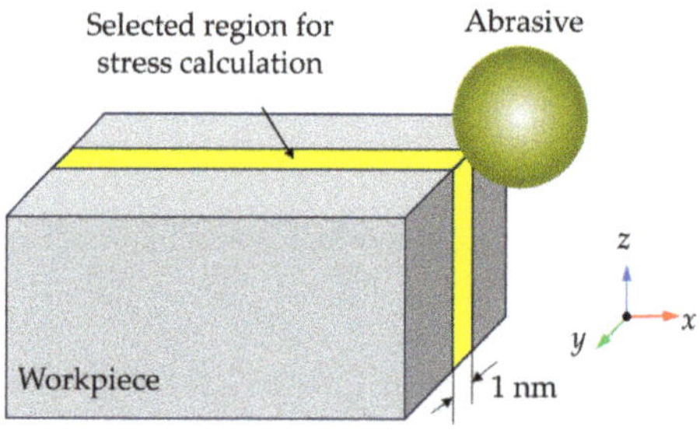

Figure 12. Illustration of the region selected for stress calculation.

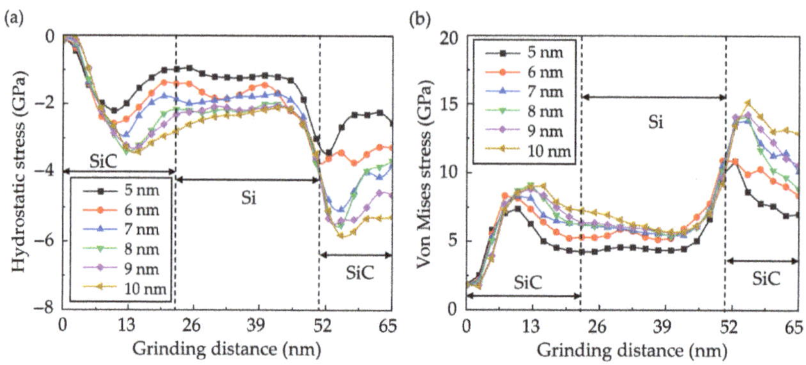

Figure 13. The average (**a**) hydrostatic stress and (**b**) von Mises stress of the selected region during nano-grinding with abrasives of different sizes.

4. Conclusions

In the current work, the effect of abrasive size on the machining mechanism of RB-SiC during nano-grinding was investigated using MD simulation. The surface morphology and subsurface deformation mechanism were investigated. Although the simulation scale is much smaller than that in experiments, the results of this simulation provide an atomic insight into the couple defamation behavior of SiC and Si phases, which is important for surface integrity during nanoscale/macroscale machining of RB-SiC. The following is a summary of the main conclusions:

1. When small abrasives are used, surface swelling of SiC is observed due to the bending and tearing of SiC at the interfaces. As the abrasive radius increases, the surface swelling is mainly formed by Si atoms, which is mainly identified as elastic recovery. The surface roughness shows an increasing trend as the abrasive size increases, indicating the significance of elastic recovery on the overall surface roughness.
2. During nano-grinding, Si has a higher material removal rate than SiC. As the abrasive size increases, the material removal rate gradually decreases, and the depth of plastic deformation is obviously increased. Stocking of Si is more apparent at the interface, and more plastic deformation patterns are observed inside the SiC grains. Further-

more, an obvious sliding of the grains is observed, forming edge cracks at the margin of the workpiece.

3. In the subsurface workpiece, the disordered Si atoms can be stocked and squeezed into the phase boundaries. When large abrasives are used, the HPPT of Si is promoted, and the squeeze of the disordered Si becomes more obvious. More disordered Si atoms are generated, while variation in the disordered SiC atoms is slight.
4. As the abrasive size increases, compression of the workpiece material is significantly enhanced, and more dislocations are formed in the subsurface of the workpiece. The bending of SiC becomes more apparent with an obvious increase in von Mises stress when machining from SiC to Si. Meanwhile, the von Mises stress can be increased by the enhanced extrusion of Si as the abrasive moves from Si to SiC.

Author Contributions: Conceptualization, investigation, and writing—original draft preparation, H.M.; methodology, and software, H.M. and X.C. (Xie Chen); validation, C.L. and X.C. (Xiaojiang Cai); formal analysis, X.C. (Xie Chen); data curation, H.M. and X.C. (Xiaojiang Cai); supervision and writing—review and editing, X.C. (Xiaojiang Cai). All authors have read and agreed to the published version of the manuscript.

Funding: This research received no external funding.

Data Availability Statement: The original contributions presented in this study are included in the article. Further inquiries can be directed to the corresponding author.

Conflicts of Interest: The authors declare no conflicts of interest.

References

1. Zhang, X.; Hu, H.; Wang, X.; Luo, X.; Zhang, G.; Zhao, W.; Wang, X.; Liu, Z.; Xiong, L.; Qi, E.; et al. Challenges and strategies in high-accuracy manufacturing of the world's largest SiC aspheric mirror. *Light Sci. Appl.* **2022**, *11*, 310. [CrossRef] [PubMed]
2. Wang, R.; Cheung, C.F.; Zang, Y.; Wang, C.; Liu, C. Material removal rate optimization with bayesian optimized differential evolution based on deep learning in robotic polishing. *J. Manuf. Syst.* **2025**, *78*, 178–186. [CrossRef]
3. Wang, H.; Dong, Z.; Kang, R.; Gao, S. Surface characteristics and material removal mechanisms during nano-grinding on C-face and Si-face of 4H-SiC crystals: Experimental and molecular dynamics insights. *Appl. Surf. Sci.* **2024**, *665*, 160293. [CrossRef]
4. Liu, C.; Chu, J.; Zhang, J.; Zhang, J.; Chen, X.; Xiao, J.; Xu, J. Effect of tool rake angle on the material removal mechanism transition of single-crystal silicon: A molecular dynamics study. *Int. J. Adv. Manuf. Technol.* **2021**, *115*, 3631–3644. [CrossRef]
5. Liu, C.; Chen, X.; Ke, J.; She, Z.; Zhang, J.; Xiao, J.; Xu, J. Numerical investigation on subsurface damage in nanometric cutting of single-crystal silicon at elevated temperatures. *J. Manuf. Process.* **2021**, *68*, 1060–1071. [CrossRef]
6. Chavoshi, S.Z.; Xu, S.; Luo, X. Dislocation-mediated plasticity in silicon during nanometric cutting: A molecular dynamics simulation study. *Mater. Sci. Semicond. Process.* **2016**, *51*, 60–70. [CrossRef]
7. Liu, C.; Xu, W.; Zhang, J.; Xiao, J.; Chen, X.; Xu, J. Numerical investigation on the temperature effect in nanometric cutting of polycrystalline silicon. *Int. J. Mech. Sci.* **2022**, *220*, 107172. [CrossRef]
8. Geng, R.; Yang, X.; Xie, Q.; Zhang, R.; Zhang, W.; Qiu, H.; Mu, R.; Yang, W.; Li, R. Investigation of surface integrity and material removal mechanism of single-crystal Si subjected to micro-laser-assisted machining. *Infrared Phys. Technol.* **2021**, *118*, 103868. [CrossRef]
9. Liu, C.; Chen, X.; Zhang, J.; Zhang, J.; Chu, J.; Xiao, J.; Xu, J. Molecular dynamic simulation of tool groove wear in nanoscale cutting of silicon. *AIP Adv.* **2020**, *10*, 015327. [CrossRef]
10. Wang, J.; Zhang, X.; Fang, F.; Chen, R. A numerical study on the material removal and phase transformation in the nanometric cutting of silicon. *Appl. Surf. Sci.* **2018**, *455*, 608–615. [CrossRef]
11. Meng, B.; Yuan, D.; Xu, S. Study on strain rate and heat effect on the removal mechanism of SiC during nano-scratching process by molecular dynamics simulation. *Int. J. Mech. Sci.* **2019**, *151*, 724–732. [CrossRef]
12. Sahoo, S.; Mannan, S.; Tiwari, U.; Kaware, R.R.; Ye, Z.; Gosvami, N.N.; Krishnan, N.M.A. Atomistic insights into scratch-induced structural evolution of silica glass. *Acta Mater.* **2025**, *282*, 120459. [CrossRef]
13. Bhamra, J.S.; Ewen, J.P.; Ayestarán Latorre, C.; Bomidi, J.A.R.; Bird, M.W.; Dini, D. Atomic-scale insights into the tribochemical wear of diamond on quartz surfaces. *Appl. Surf. Sci.* **2023**, *639*, 158152. [CrossRef]
14. Pradhan, T.; Pal, S.; Deng, C. Investigation on wear-resistance of nanocrystalline Pt-Au by molecular dynamics simulations. *Tribol. Int.* **2023**, *189*, 108966. [CrossRef]

15. Goel, S.; Kovalchenko, A.; Stukowski, A.; Cross, G. Influence of microstructure on the cutting behaviour of silicon. *Acta Mate* **2016**, *105*, 464–478. [CrossRef]
16. Liu, Y.; Li, B.; Kong, L. Molecular dynamics simulation of silicon carbide nanoscale material removal behavior. *Ceram. Int.* **2018** *44*, 11910–11913. [CrossRef]
17. Meng, B.; Qiu, P.; Yuan, D.; Xu, S. Influence of microstructure on the diamond-machinability of hot-pressed silicon carbide: A molecular dynamics study. *Ceram. Int.* **2019**, *45*, 22872–22879. [CrossRef]
18. Liu, C.; Ke, J.; Yin, T.; Yip, W.S.; Zhang, J.; To, S.; Xu, J. Cutting mechanism of reaction-bonded silicon carbide in laser-assisted ultra-precision machining. *Int. J. Mach. Tools Manuf.* **2024**, *203*, 104219. [CrossRef]
19. Liu, C.; Zhang, J.; Zhang, J.; Chu, J.; Chen, X.; Xiao, J.; Xu, J. Numerical investigation on material removal mechanism in elliptical vibration cutting of single-crystal silicon. *Mater. Sci. Semicond. Process.* **2021**, *134*, 106019. [CrossRef]
20. Ameli Kalkhoran, S.N.; Vahdati, M.; Yan, J. Effect of relative tool sharpness on subsurface damage and material recovery in nanometric cutting of mono-crystalline silicon: A molecular dynamics approach. *Mater. Sci. Semicond. Process.* **2020**, *108*, 104868 [CrossRef]
21. Voronoi, G. Nouvelles applications des paramètres continus à la théorie des formes quadratiques. Deuxième mémoire. Recherches sur les parallélloèdres primitifs. *J. Für Die Reine Und Angew. Math. (Crelles J.)* **1908**, *134*, 198–287. [CrossRef]
22. Hirel, P. Atomsk: A tool for manipulating and converting atomic data files. *Comput. Phys. Commun.* **2015**, *197*, 212–219. [CrossRef
23. Goel, S.; Luo, X.; Agrawal, A.; Reuben, R.L. Diamond machining of silicon: A review of advances in molecular dynamics simulation. *Int. J. Mach. Tools Manuf.* **2015**, *88*, 131–164. [CrossRef]
24. Erhart, P.; Albe, K. Analytical potential for atomistic simulations of silicon, carbon, and silicon carbide. *Phys. Rev. B* **2005**, *71* 035211. [CrossRef]
25. Liu, C.; Zhang, J.; Zhang, J.; Chen, X.; Xiao, J.; Xu, J. A simulation investigation on elliptical vibration cutting of single-crysta silicon. *Int. J. Adv. Manuf. Technol.* **2020**, *108*, 2231–2243. [CrossRef]
26. Liu, C.; He, W.; Chu, J.; Zhang, J.; Chen, X.; Xiao, J.; Xu, J. Molecular Dynamics Simulation on Cutting Mechanism in the Hybrid Machining Process of Single-Crystal Silicon. *Nanoscale Res. Lett.* **2021**, *16*, 66. [CrossRef]
27. Chavoshi, S.Z.; Luo, X. Molecular dynamics simulation study of deformation mechanisms in 3C–SiC during nanometric cutting at elevated temperatures. *Mater. Sci. Eng. A* **2016**, *654*, 400–417. [CrossRef]
28. Chavoshi, S.Z.; Goel, S.; Luo, X. Influence of temperature on the anisotropic cutting behaviour of single crystal silicon: A molecular dynamics simulation investigation. *J. Manuf. Process.* **2016**, *23*, 201–210. [CrossRef]
29. Plimpton, S. Fast parallel algorithms for short-range molecular dynamics. *J. Comput. Phys.* **1995**, *117*, 1–19. [CrossRef]
30. Stukowski, A. Visualization and analysis of atomistic simulation data with OVITO–the Open Visualization Tool. *Model. Simul. Mater. Sci. Eng.* **2010**, *18*, 015012. [CrossRef]
31. Chen, J.; Liu, C.; Liu, H.; Zhang, B.; To, S. Atomic insight into the speed effect on deformation mechanisms in nano-scratching of monocrystalline iron. *Precis. Eng.* **2024**, *92*, 219–230. [CrossRef]
32. Liu, C.; Zhang, J.; Zhang, J.; Chen, X.; He, W.; Xiao, J.; Xu, J. Influence of micro grooves of diamond tool on silicon cutting: A molecular dynamic study. *Mol. Simul.* **2020**, *46*, 92–101. [CrossRef]
33. Chen, X.; Chu, J.; Zhu, Z.; Liu, C. Effect of temperature on surface morphology of single-crystal silicon in nanometric cutting. *Appl. Surf. Sci.* **2025**, *684*, 161957. [CrossRef]
34. Maras, E.; Trushin, O.; Stukowski, A.; Ala-Nissila, T.; Jónsson, H. Global transition path search for dislocation formation in Ge on Si(001). *Comput. Phys. Commun.* **2016**, *205*, 13–21. [CrossRef]
35. Dai, H.; Zhou, Y.; Zhang, F. Atomistic simulation of influence of laser nano-structured diamond abrasive on the polishing behavior of silicon. *Mater. Sci. Semicond. Process.* **2020**, *105*, 104706. [CrossRef]

micromachines

rticle

;urface Integrity of Glass-Ceramics by Laser-Assisted)iamond Cutting

.awei Li , Fang Ji * and Feifei Xu

Institute of Machinery Manufacturing Technology, China Academy of Engineering Physics, Mianyang 621900, China
* Correspondence: jfang2004@sina.com

Abstract: Glass-ceramic optical components are extensively employed in advanced optical systems. The high-hardness and low-fracture toughness of glass-ceramics make it prone to cracks and subsurface damage during conventional cutting. The laser-assisted diamond cutting method can significantly improve the nano-cutting performance of glass-ceramics by locally heating and softening the material. However, its dynamic removal mechanisms remain unclear. The coupling mechanisms between the laser thermal field and the mechanical response of the material require further investigation. This study aims to reveal the dynamic removal mechanisms of glass-ceramics under laser-assisted nanoscale cutting conditions through numerical simulations and systematic experiments. It includes a systematic analysis of the effects of laser heating on chip morphology, temperature fields, stress fields, and cutting forces using a laser-assisted nano-cutting model. Additionally, through nanoscale taper cutting experiments, this study quantifies the enhancement effect of laser power on the critical depth of no observed surface cracks (NOSC). Finally, subsurface integrity results elucidate the mechanisms through which laser assistance inhibits crack propagation. The findings will provide theoretical support for optimizing laser-assisted cutting parameters and achieving high-quality machining of glass-ceramics.

Keywords: glass-ceramics; removal mechanism; laser-assisted diamond cutting; surface integrity

1. Introduction

Glass-ceramics exhibit excellent properties such as a near-zero thermal expansion coefficient and high transparency across a wide spectral range [1], which has led to their widespread application in fields such as aerospace observation and optical instruments. Glass-ceramics possess good mechanical strength and hardness, while their fracture toughness is relatively low, indicating they're a typical hard and brittle material [2]. High-quality machining is challenging because brittle removal often occurs during machining, leading to surface and subsurface damage that affects the overall performance of the component [3].

The performance and stability of glass-ceramic optics are highly dependent on the finish quality of optical components. At present, the form accuracy of glass-ceramic optics is required to meet the specification of 1/30–1/50λ. To meet the required shape accuracy and surface quality control, glass-ceramic optical components are machined through multi-processes, including precision grinding, rapid polishing, rough polishing, and fine polishing. Precision grinding can achieve sub-micron surface accuracy, and is widely used in the manufacture of various optical materials [4]. However, grinding tends to leave severe subsurface damage, significantly limiting the improvement of the machining efficiency for glass-ceramic optical components. In addition, when it comes to

 https://doi.org/10.3390/mi16091054

special structural components such as microlens arrays and concave cones, grinding wheel interference greatly limits the application of this method.

Ultra-Precision turning Technology has gained extensive application [5], owing to its high cost-effectiveness, environmentally friendly process, superior form accuracy control capabilities, and exceptional adaptability to complex geometries [6,7]. However, its cutting performance on hard, brittle materials is constrained by critical edge degradation issues. Specifically, acute diamond tool edges incur micro-chipping, fracture, and accelerated mechanical wear [8] during machining processes. This damage mechanism fundamentally limits practical implementation in precision optics manufacturing [9]. Zhang et al. [10] observed severe micro-groove mechanical wear on diamond tools when turning RB-SiC material. Conventional diamond turning faces significant challenges in achieving ductile regime removal of fused silica, as hard, brittle materials inherently undergo brittle fracture under standard machining conditions. Scratch testing at a 30 nm cutting depth confirmed persistent brittle removal characteristics in fused silica, with subsurface cracks extending up to 200 nm [11]. This critical crack-to-cut depth ratio conclusively demonstrates the limitations of conventional SPDT for ductile processing of this optical material. At present, there is no mature technology application for the mass manufacturing of glass-ceramic optical components through diamond turning.

Laser-assisted diamond turning (LADT) guides the laser beam to pass through a transparent diamond tool to locally heat the workpiece material in the cutting region [12]. This method effectively enhances the ductile-brittle transition depth [13], facilitating ductile ductile removal of hard and brittle materials [14], and achieves a higher material removal rate compared to ultraprecision grinding while maintaining the same surface quality. The feasibility of the LADT method has been successfully demonstrated in sapphire [15], silicon [16], and tungsten carbide [17] materials before. The surface roughness achieved by LADT can reach 1–5 nm, showing its broad application prospects.

At present, there is a strong demand for high-quality machining of glass-ceramic, and there is an urgent need to improve machining efficiency and quality by turning instead of grinding. Glass-ceramic material presents an obvious decrease in hardness at elevated temperature, which has been demonstrated by Milhans et al. [18] using high-temperature nanoindentation techniques. Their results clearly demonstrated that both the hardness and elastic modulus of the material decreased significantly as the testing temperature was elevated from 25 °C to 400 °C, confirming the distinct thermal softening characteristics of this glass-ceramic. This substantial reduction in hardness reflects an enhanced plastic deformation capacity of the material at elevated temperatures, proving that LADT is a highly promising solution. However, there are still several core technical bottlenecks that need to be addressed. Firstly, the laser heating thermal field needs to be accurately controlled to achieve sufficient high-temperature softening of the material and also avoid thermal damage and high-temperature graphitization of the diamond tool [17]. Furthermore, the enhancement of nano-cutting performance under laser-assisted nano-cutting conditions for glass-ceramics is still unclear, and it is crucial to determine the critical depth of no observed surface cracks (NOSC) through taper cutting experiments. Finally, the impact of laser-assisted nano-cutting on the subsurface integrity of glass-ceramics needs to be systematically analyzed to provide guidance for the material removal depth in subsequent polishing steps.

To address the research gaps and bottlenecks in glass-ceramic machining using the LADT method, we conduct a systematic investigation of the laser heating thermal field prediction and control. A laser-assisted nano-cutting model for glass-ceramics is established, and a series of experiments is performed to elucidate the material-removal mechanisms under in-situ laser assistance. These findings provide a theoretical foundation for high-quality, high-precision machining of glass-ceramic optical components.

2. Laser Power Selection

Material temperature is a vital index in laser-assisted turning (In-LAT) and significantly affects machining quality. An appropriate heating temperature reduces material hardness and improves machinability, while excessive temperature induces undesired thermal damage and severe diamond tool wear. It is essential to keep the laser heating temperature within an optimal range, which mainly depends on the irradiated laser power. Numerical calculation facilitates thermal field analysis and cuts down experimental efforts; combined with thermal initial conditions and complex boundary conditions, the workpiece thermal field can be precisely calculated.

In the case of a boundary heat source, if the workpiece moves in the positive x-direction with a constant velocity V_{laser}, the governing heat conduction equation can be defined by:

$$\nabla^2 T - \frac{\rho c_p}{k} V_{laser} \frac{\partial T}{\partial x} = 0 \tag{1}$$

where ρ is the density in kg/m^3, c_p is the material specific heat capacity in $J/(kg{\cdot}K)$, k is the thermal conductivity in $W/(m{\cdot}K)$.

The boundary condition of the boundary laser heat can be expressed as:

$$k\frac{\partial T}{\partial z}(x, y, 0) = -\frac{2AP}{\pi R^2} e^{-\frac{2(x^2+y^2)}{R^2}} \tag{2}$$

where $-\infty < x < +\infty$; $-\infty < y < +\infty$; $z > 0$, where R is the radius of the laser spot, P is the energy of the laser irradiation on the material surface, and A is the absorptivity of the material to the laser. Since the absorption coefficient of glass-ceramics for the 1064 nm laser is high, the heat flux is simplified by assuming the laser is absorbed on the material surface, loading only as a surface heat source without considering absorption in the depth direction.

The basic equation for the analysis of heat conduction is Fourier's law:

$$q_n = -K_n \frac{\partial T}{\partial n} \tag{3}$$

where heat flux q_n is the heat transfer rate in the n direction per unit area perpendicular to the direction of heat flow. $\partial T/\partial n$ (K/m) is the temperature gradient in the direction n. The K_n ($W{\cdot}m^{-1}{\cdot}K^{-1}$) is the thermal conductivity in the n direction.

The laser-induced heating temperature is a critical process parameter for LADT technology. However, accurate determination of the workpiece temperature field during LADT is impeded by the limited spatial resolution of existing thermometric devices and by the occlusion of the cutting zone by the tool. Consequently, a finite-element thermal-field model of the workpiece has been established. Considering the extremely low absorption coefficient of diamond at the working wavelength of 1064 nm and its near-ideal thermal expansion properties, the temperature variation within the diamond tool is negligible [19]. Thus, only the workpiece thermal field was calculated in the simulation model. During the simulations, the peak temperature within the workpiece was employed as a diagnostic probe to assess the evolution of the thermal field under different laser-power levels. To characterize the surface temperature distribution induced by laser irradiation at various power settings, a simplified three-dimensional transient heat-conduction model was constructed using the finite element simulation software. Because the projected shape of the laser beam is irregular and not readily amenable to analytical description, the beam was approximated by a circular spot in the finite-element model, as illustrated in Figure 1a. The laser spot in

Figure 1a deviates from a Gaussian distribution owing to the laser's high-speed scanning, forming a symmetric "comet tail"-like profile.

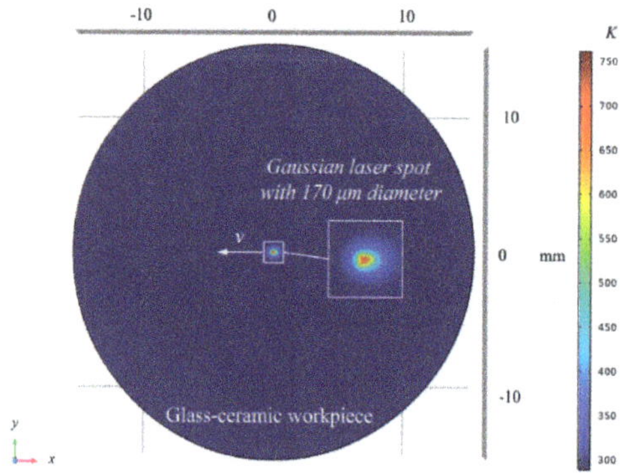

(a)

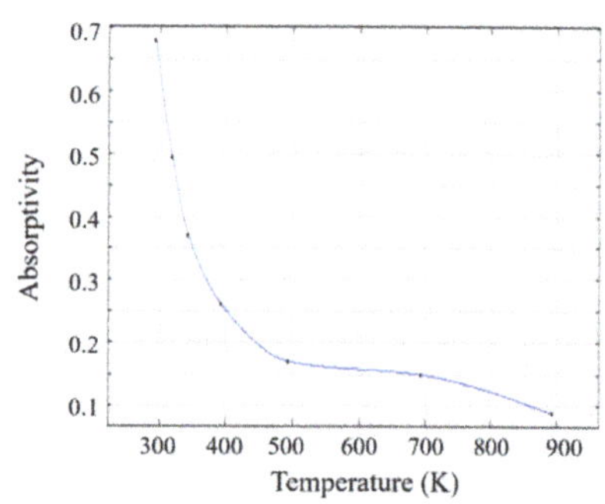

(b)

Figure 1. FEA simulation of the temperature field caused by the laser heating: (**a**) model dimensional setup, (**b**) temperature-dependent absorption parameters of the glass-ceramics.

To improve temperature simulation accuracy, a laser temperature measurement experiment was conducted. A 1064 nm laser irradiated the workpiece surface through a diamond tool, with thermocouples around the heating point. Using a micro laser-assisted system, thermocouples, and a Φ30 × 5 mm glass-ceramic workpiece, temperatures at 2.5 mm and 4 mm from the heating point under varying laser powers were measured and compared with simulations. Temperature was recorded in three stages: preheating (0–40 s), laser heating (40–160 s), and natural cooling (160–450 s). Initial simulations deviated greatly from experiments due to ignoring the temperature—dependent absorptivity of glass-ceramic (0.68 at room temperature for 1070 nm laser). Via COMSOL parameter optimization, a nonlinear absorptivity-temperature relationship (absorptivity decreases with temperature) was obtained.

The numerical simulations employed a moving surface heat source to represent a continuous-wave 1064 nm laser (spot diameter = 170 μm) operating at power levels of 10, 20, 30, and 35 W. The laser irradiated a glass-ceramic workpiece (diameter = 30 mm, thickness = 5 mm) while scanning radially at a traverse speed of 200 mm min^{-1}. The finite-element model incorporated forced convection from the cutting fluid, natural convection to ambient air (heat-transfer coefficient = 20 $W{\cdot}m^{-2}{\cdot}K^{-1}$), and radiative heat loss. A convective heat-transfer coefficient of 300 $W{\cdot}m^{-2}{\cdot}K^{-1}$ for the minimum-quantity lubrication (MQL) was adopted, following Kurgin et al. [20]. To balance computational accuracy and efficiency, mesh sizes ranging from 2 μm to 150 μm were applied on the workpiece end face. Based on experimental optimization of temperature-dependent absorption parameters of the glass-ceramics, an interpolation function was established to define the material's absorptivity as a function of temperature, as illustrated in Figure 1b. Through experimental verification, the errors of the optimized model under different heating powers are within 7.2% at 2.5 mm from the center and within 5.5% at 4 mm from the center. These error ranges are within the acceptable limits for experimental measurements, thereby validating the rationality and reliability of the optimized temperature field model. The simulation parameters have been listed in Table 1. The physical properties of the glass-ceramics were referenced from [21].

Table 1. Thermal field simulation parameters.

Parameters	Values
Workpiece Material	Glass-ceramics
Laser Power (P)	10, 20, 30, 35 W
Laser Beam Diameter	170 μm
Scanning Speed	200 mm/min
Absorptivity	Temp-dependent
Material Emissivity (ε)	0.62

The maximum temperatures of the workpiece under laser powers of 10 W, 20 W, 30 W, and 35 W are shown in Figure 2. Figure 2 exhibits fluctuations attributed to mesh size limitations in the finite element simulation, where the more fluctuating curve corresponds to raw data extracted from the model and the smoother one denotes the fitted version. As observed from the temperature curves in Figure 2, the workpiece temperature increases significantly with rising laser power from 10 W to 35 W. In all cases, a rapid temperature rise occurs during the initial heating stage, followed by a stabilization phase at different temperature levels. For instance, a low laser power of 10 W stabilizes around 550 K, while a high power of 35 W results in a steady-state temperature of approximately 1280 K. The higher laser power delivers greater energy density, allowing the workpiece surface to rapidly absorb heat and increase in temperature within a short time. Given that the onset of diamond graphitization is generally accepted to occur near 1000 K—with further acceleration at higher temperatures [22]—it is noteworthy that a laser power of 30 W yields a steady-state surface temperature of approximately 1000 K. At this temperature, the risk of diamond graphitization is mitigated, while the laser-induced softening of the glass-ceramics surface is maximized. This reduction in mechanical strength facilitates material removal by the diamond tool with minimal cutting force.

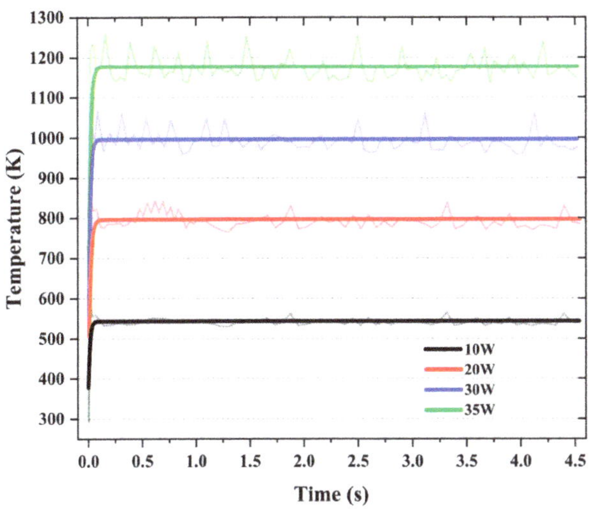

Figure 2. Simulated maximum temperature of the glass-ceramics workpiece under different laser power conditions.

3. Experimental Setup

The NOSC depth is a key metric for evaluating the machining performance of hard and brittle materials. This work systematically examines the effect of laser in-situ heating on the NOSC depth of glass-ceramics by conducting taper cutting experiments under various laser-power conditions. Experimental validation was performed on an ultra-precision lathe using a cylindrical glass-ceramic workpiece (Ø 30 mm, 5 mm thick). To eliminate the influence of the initial surface condition, all specimens were meticulously polished to guarantee Sa < 2 nm, resulting in a near-defect-free surface. The workpiece was mounted on the spindle end face with a precision fixture offering fine adjustment. Before each cut, the fixture was calibrated in situ with an integrated micrometer to guarantee that the workpiece end surface was exactly perpendicular to the spindle axis. The experiment was carried out with a single-crystal diamond tool (rake angle −35°, clearance angle 10°, nose radius 0.5 mm) mounted in a custom tool holder that maintained a vertical tool face throughout the process.

In the experiments, laser power was set at 0 W, 10 W, 15 W, 20 W, 25 W, and 28 W. Although it was found through simulation in Section 2 that a 30 W laser is optimal, the laser used in the experiment supports a maximum laser output power of 75 W. When the laser spot is adjusted to the ideal position, the maximum laser power emitted from the tool is 28 W. Therefore, the maximum laser power set in the experiment is 28 W. The taper cutting motion in the diamond cutting experiments was achieved through precise interpolation of the machine's *X*-axis and *Z*-axis, as shown in Figure 3, allowing a preset constant slope ratio of 1:1000. Furthermore, by rotating the spindle's C-axis, multiple independent taper cutting tests could be performed at different angular positions while the workpiece remained in the same clamping state. To ensure the reliability of the experimental results and exclude influences from random factors, four independent taper cutting experiments were repeated for each specified laser power level, and the final analysis utilized the average values from these four repeated experiments. Specific cutting process parameter settings are detailed in Table 2.

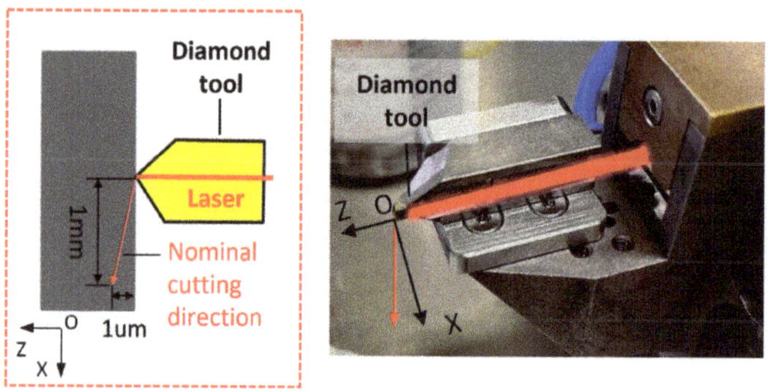

Figure 3. Schematic diagram of glass-ceramics taper cutting experiment.

Table 2. Parameters for taper cutting experiments.

Parameters	Values/Descriptions
Tool Parameters	−35° rake angle, 10° clearance angle, R0.5 mm
Cutting Speed	200 mm/min
Laser Power	0 W, 10 W, 15 W, 20 W, 25 W, 28 W
Sample Size	Φ30 × 5 mm
Slope Ratio	1:1000

4. Critical Depth of No Observed Surface Cracks

To assess the NOSC depth of glass-ceramics under different parameter groups, the white light interferometer was employed to measure the surface morphology after taper cutting. The experimental results indicate that laser power significantly enhanced the

NOSC depth of glass-ceramics. The exact positions for extracting the cross-sectional profiles in Figure 4 are marked with blue dashed lines. Without laser assistance (0 W), the NOSC depth was only 36.519 nm (Figure 4a). When the laser power increased to 10 W, the transition depth significantly increased to 100.264 nm (Figure 4b). And at 28 W power conditions, it further increased to 139.444 nm (Figure 4f). This change indicates that the material was heated and softened by the in-situ laser heating, reducing the inter-grain bonding forces and allowing glass-ceramics to undergo plastic deformation more easily, thereby enhancing machining performance.

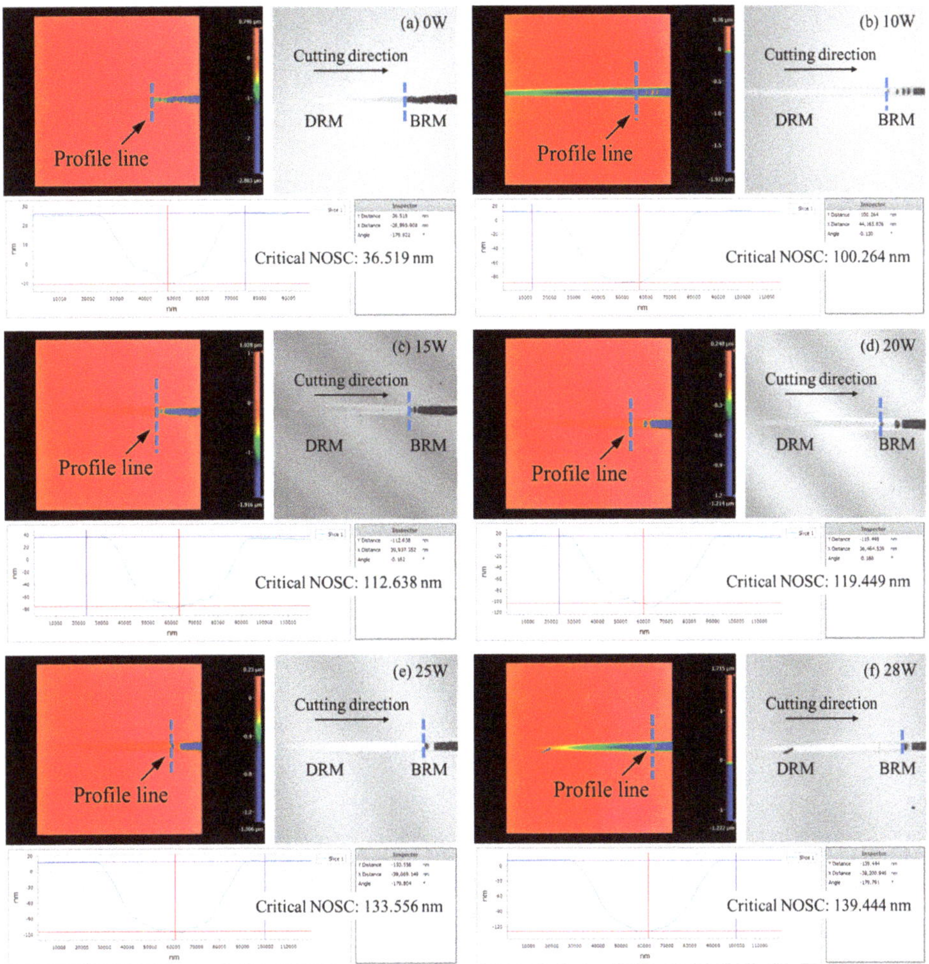

Figure 4. Critical NOSC depth of grass-ceramics under laser assistance with laser power of: (**a**) 0 W; (**b**) 10 W; (**c**) 15 W; (**d**) 20 W; (**e**) 25 W; (**f**) 28 W.

Based on the experimental results (Table 3), the critical NOSC depth increases with laser power, exhibiting a monotonic rise up to 20 W. Beyond this threshold, the growth asymptotically approaches saturation. The statistical distribution of NOSC depths (Figure 5) demonstrates a high degree of repeatability and consistent transition depths across the various laser-power settings.

Table 3. Experimental data of ductile-brittle transition depth as a function of laser power.

Power (W)	Experimental Results1 (nm)	Experimental Results2 (nm)	Experimental Results3 (nm)	Experimental Results4 (nm)
0	36.519	41.197	41.825	38.321
10	93.14	102.12	91.91	100.264
15	112.638	111.356	143.609	114.651
20	119.449	143.437	112.127	114.192
25	133.556	131.092	136.456	128.151
28	129.93	139.444	134.945	131.635

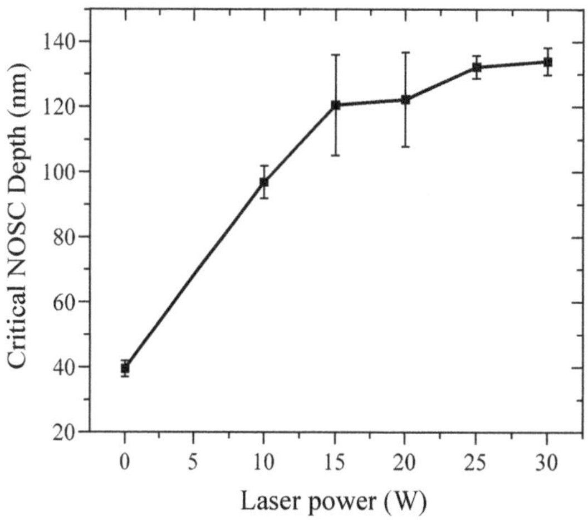

Figure 5. Critical NOSC depth from taper cutting experiments at different laser powers.

Combining the experimental results, this study not only reveals the significant enhancement of critical NOSC depth of glass-ceramics due to laser-assisted cutting, but also provides a theoretical basis for optimizing LADT machining parameters. In practical applications, suitable power ranges and machining parameters can be selected based on the matching relationship between undeformed chip thickness and laser power to achieve high-efficiency and high-quality machining of glass-ceramics.

5. Surface Generation Mechanism

5.1. Numerical Simulation

The laser-assisted nano-cutting model was established based on the finite element analysis (FEA) to evaluate the influence of laser in-situ heating on the glass-ceramics material removal mechanism. The chip temperature distribution, cutting forces, and stress fields were systematically calculated and analyzed.

To achieve an appropriate compromise between computational accuracy and efficiency, a two-dimensional orthogonal-cutting finite-element model was developed in this study. The geometric configuration of this model is shown in Figure 6. Laser heating was introduced via user-defined subroutines written in Fortran, which impose a time-dependent thermal load synchronized with the instantaneous position of the cutting tool, thereby reproducing the characteristics of in-situ laser heating. To maximize heating efficiency and tightly control the thermal-affected zone, the laser-spot centre was continuously positioned at the tool tip. Accordingly, dynamic heat-flux boundary conditions were applied to the workpiece surface, with the heat-flux density calculated from the Gaussian-beam energy distribution for a beam of 170 μm diameter.

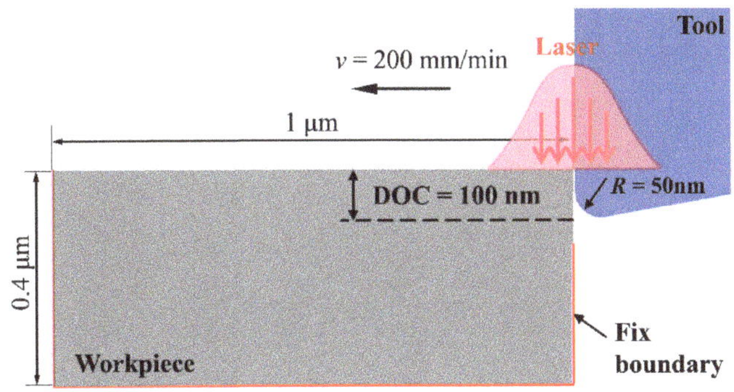

Figure 6. Model schematic of laser-assisted cutting of glass-ceramics.

The geometric dimensions and cutting parameters employed in the numerical model were chosen to replicate the experimental conditions. The workpiece is defined with macroscopic dimensions of L = 1.0 μm (length) and H = 0.4 μm (height). The cutting tool is made of monocrystalline diamond; owing to its exceptionally high hardness and elastic modulus, the tool is modeled as an analytically rigid body. The tool geometry is specified as a rake angle of 0°, a clearance angle of 10°, and, to account for the finite sharpness of a real diamond edge, an edge radius of 50 nm is introduced. Cutting conditions identical to those used in the experiments are imposed: a nominal depth of cut of 100 nm and a cutting speed of 200 mm min^{-1}. During the simulation, the tool translates uniformly in the negative X-direction. A complete list of the model parameters is provided in Table 4.

Table 4. Simulation parameters for the laser-assisted cutting finite element model.

Parameters	Values
Workpiece Material	Glass-ceramics
Cutting Depth	100 nm
Tool Edge Radius	50 nm
Cutting Speed	200 mm/min
Minimum Grid Cell Size	5 nm
Laser Power (P)	0 W, 28 W
Laser Beam Diameter	170 μm
Material Emissivity (ε)	0.62
Convection Thermal Conductivity Coefficient (h)	300 W/(m^2·K)

5.1.1. Thermal Field

Using a two-dimensional orthogonal cutting model, the temperature field of the glass-ceramic workpiece and cutting chips was investigated under conventional and 28 W laser-assisted cutting conditions. Figure 7 presents the comparison at a cutting distance of 0.87 μm. The simulation results clearly show that laser assistance markedly modifies the cutting process. In conventional nano-cutting (Figure 7a), material removal occurs in brittle mode, producing a segmented, discontinuous chip. The temperature rise in the workpiece is dominated by tool–workpiece friction, and the elevated-temperature region is confined to the primary and secondary deformation zones, reaching a maximum of 874 K. When applying laser-assisted cutting (Figure 7b), the laser energy fundamentally changes the material's removal behavior. The chip becomes noticeably more curled and continuous, which indicates a transition from brittle fracture to ductile removal mode. Under the 28 W laser power, the laser heating becomes the dominant thermal source, significantly raising the temperature in the cutting zone. The maximum temperature reaches 1094 K, and this localized softening of the glass-ceramics directly contributes to the improved ductility and smoother chip formation.

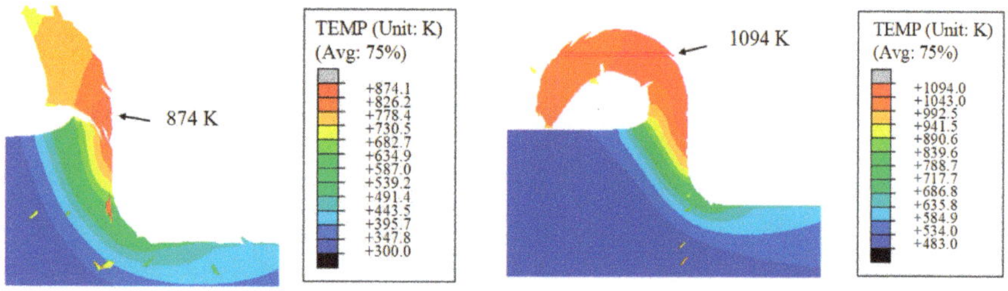

(**a**) (**b**)

Figure 7. Workpiece thermal field in the process of: (**a**) conventional nano-cutting; (**b**) laser-assisted nano-cutting.

5.1.2. Cutting Force

Cutting force is a fundamental physical quantity in ultra-precision machining and serves as a critical metric for assessing tool load, predicting wear progression, and optimizing machining parameters. In the FEA simulations, the forces acting on the tool were recorded continuously with a temporal resolution of 1 ns. To extract the primary trend of cutting-force variations and suppress high-frequency numerical noise, the raw data were post-processed using a low-pass filter based on the Fast Fourier Transform (FFT), with a cutoff frequency set at 1.25 MHz.

Figure 8 illustrates the variation of the cutting-force components as a function of cutting distance, both with and without laser heating assistance. Figure 8a shows the main (tangential) cutting force, which is aligned with the cutting-speed direction, whereas Figure 8b presents the normal (thrust) force acting perpendicular to the machined surface. The curves clearly indicate that laser heating softens the glass-ceramic material, thereby reducing its cutting resistance.

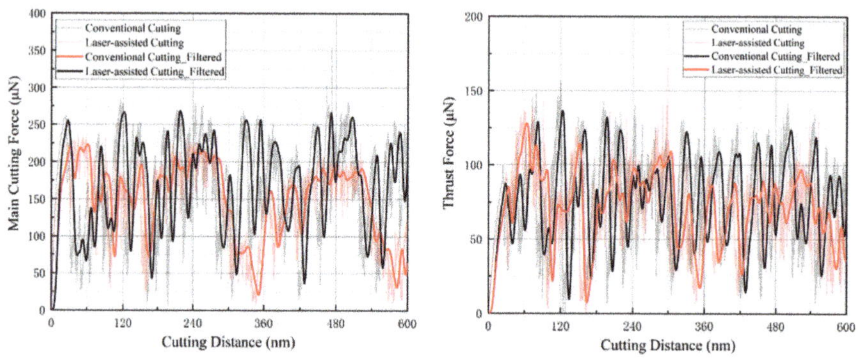

(**a**) (**b**)

Figure 8. Comparison of nano-cutting forces with and without laser assistance: (**a**) main cutting force; (**b**) normal cutting force.

To quantify this effect, the filtered force signals were averaged over the entire cutting distance. For the main force, the mean value without laser assistance is ≈195 μN, which drops to ≈160 μN when laser heating is applied—a reduction of roughly 17.9%. For the normal force, the average decreases from ≈95 μN in the conventional case to ≈85 μN under laser-assisted conditions, corresponding to a reduction of about 10.5%. These results demonstrate that LADT can markedly reduce cutting forces, confirming that laser heating induces a softening effect of glass-ceramics, thereby mitigating tool wear and extending tool life.

5.2. Subsurface Crack

To assess the subsurface quality of the machined glass-ceramic workpiece, thin samples were harvested from taper cutting representative zones for transmission electron microscopy (TEM) analysis. Using a focused ion-beam (FIB) system, an 8 × 5 × 1 μm sample was lifted from the cross-section of the ductile-brittle transition region and subsequently thinned to a uniform thickness of 50–70 nm to enable detailed TEM characterization of sub-surface characteristics. The ductile–brittle transition zone of the material is defined by the simultaneous presence of surface crack and ductile cutting features. Consequently, a localized sub-surface region of approximately 3 μm surrounding the transition point was selected for sample preparation, allowing high-resolution observation of subsurface cracks in the glass-ceramic. The sampling location is indicated in Figure 9.

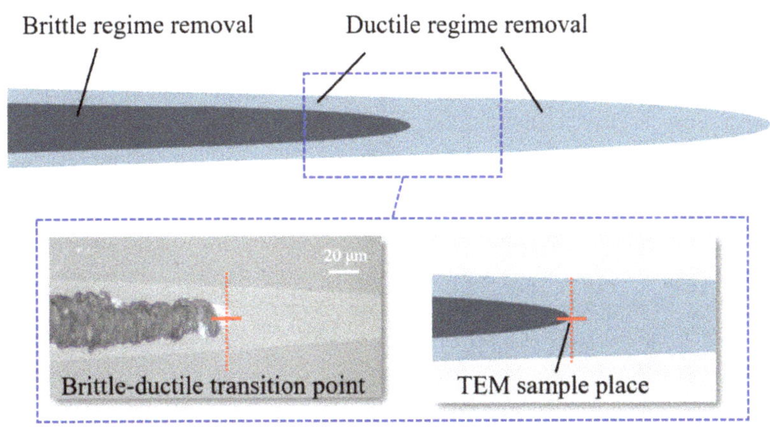

Figure 9. Schematic diagram of TEM sample position.

The exact positions for extracting the cross-sectional profiles in Figure 9 are marked with red dashed lines. Subsurface damage in glass-ceramic workpieces—manifested by crack initiation and propagation—can degrade the component's functional performance. Among the process parameters, laser power is the dominant factor that governs both the extent of crack propagation and the depth of subsurface damage. Accordingly, this subsection presents a systematic investigation of specimens machined at laser powers of 0 W, 10 W, 20 W, and 28 W (Figure 10). Surface crack-free regions are framed in the figure, permitting a direct comparison of crack-propagation length and subsurface-damage depth under the different laser-power conditions.

Under conventional nano-cutting (laser power = 0 W), the subsurface region of the glass-ceramic exhibits a dense network of interlaced cracks both in the brittle removal regime and the surface crack-free region, extending to a depth of approximately 2 μm, as shown in Figure 10a. This observation clearly demonstrates the pronounced brittleness of the glass-ceramic under traditional cutting: cracks readily propagate along grain boundaries or pre-existing defects, forming a network-like pattern.

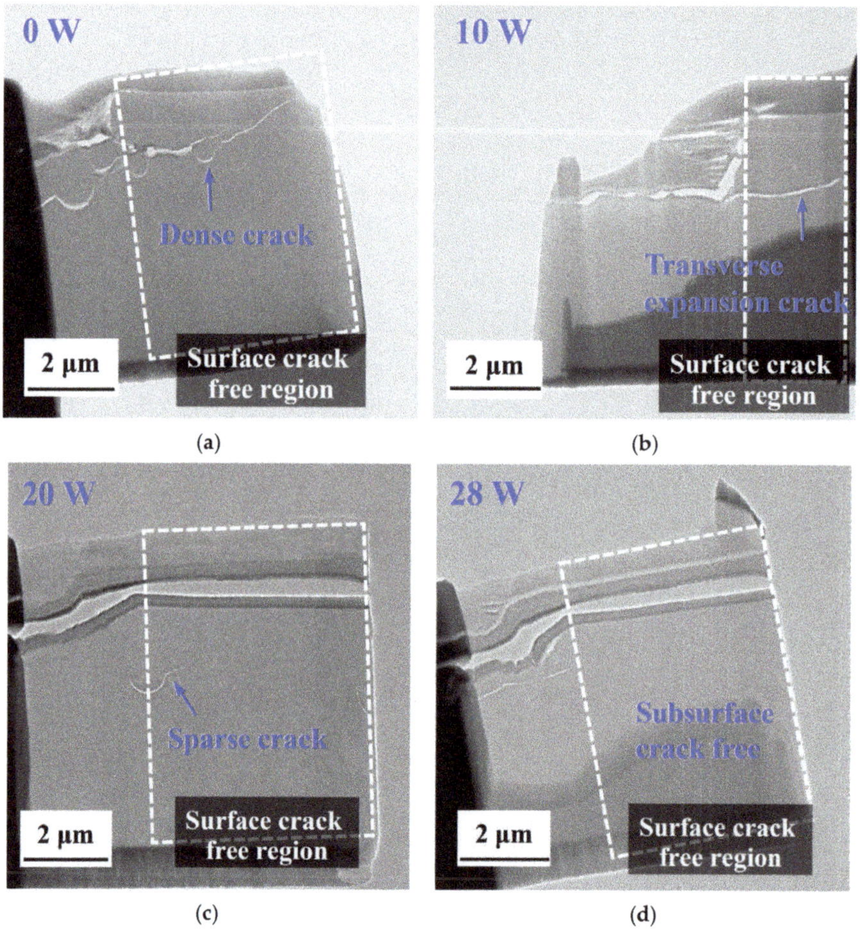

Figure 10. Comparison of subsurface damage of glass-ceramics by: (**a**) traditional Cutting; (**b**) 10 W laser-assisted nano cutting; (**c**) 20 W laser-assisted nano cutting; (**d**) 28 W laser-assisted nano cutting.

The introduction of laser energy locally thermal softens the material, thereby diminishing the driving force for crack propagation. The emergence of transversely oriented cracks indicates a change in the propagation direction. Laser-induced thermal stresses, superimposed on the pre-existing mechanical stresses, generate specific directional stresses within the glass-ceramic. Consequently, thermal softening reduces the stress concentration at crack tips, suppresses longitudinal branching, and promotes lateral crack growth as illustrated in Figure 10b. Simultaneously, residual thermal stresses combined with mechanical loading drive deeper crack penetration.

When the laser power is raised to 20 W (relative to the 10 W condition), the subsurface cracks in the surface crack-free region become sparse, with a reduced depth of approximately 3.8 μm (Figure 10c). Further increase of the laser power expands the absorbed energy, enlarging the thermal influence zone. This enlarges the plastic-deformation region during machining, reduces the proportion of brittle fracture, and curtails extensive crack propagation, leaving only a few isolated cracks.

No subsurface cracks are observed in the surface crack-free region for the 28 W laser-assisted nano cutting, as shown in Figure 10d. At this power level, the laser markedly increases the material's fracture toughness while decreasing its hardness; the lower cutting force diminishes the cutting stress, and pronounced thermal softening enhances the material's machinability. These combined effects prevent crack initiation and propagation, thereby virtually eliminating subsurface cracks.

5.3. Subsurface Integrity

Subsurface damage encompasses more than the mere presence or absence of cracks; it also involves internal microstructural alterations—such as lattice distortion, grain refinement, and microcrack initiation—that directly affect critical performance metrics, including fatigue life, wear resistance, and corrosion resistance. Consequently, a comprehensive investigation of the morphology and mechanisms of subsurface damage across various laser-power levels is required to systematically characterize the specific damage forms under each condition and to elucidate their underlying physicochemical mechanisms, thereby providing robust scientific evidence for practical engineering applications.

The subsurface damage induced by taper cutting of glass-ceramics, both without and with laser assistance, is illustrated in Figures 11 and 12, respectively. In conventional nano-cutting (Figure 11a), the dominant damage mode in the near-surface region is the formation of a distinct amorphous layer. This layer originates from intense mechanical shear that disrupts the ordered β-SiO_2 crystal lattice, yielding a disordered amorphous structure. Such a phenomenon directly reflects brittle removal, because the material lacks thermal softening and cannot accommodate appreciable ductile deformation. This observation is consistent with the shallow critical NOSC depth of 36.5 nm measured after conventional taper cutting, indicating that only an extremely thin surface region may undergo limited plastic deformation.

In contrast, the near-surface region of 28 W laser-assisted nano-cutting (Figure 12a) exhibits a uniform plastic-damage layer. Under shear stress, the crystals preferentially undergo orderly slip rather than fracture, producing a shallow plastic layer of approximately 200 nm thickness that can be readily removed by subsequent polishing. This observation is consistent with the increased critical NOSC depth observed under the 28-W laser taper cutting condition, confirming the enhanced plastic-removal capability.

(a) (b)

Figure 11. Subsurface damage of glass-ceramics under conventional nano-cutting: (**a**) near surface region; (**b**) subsurface crack region.

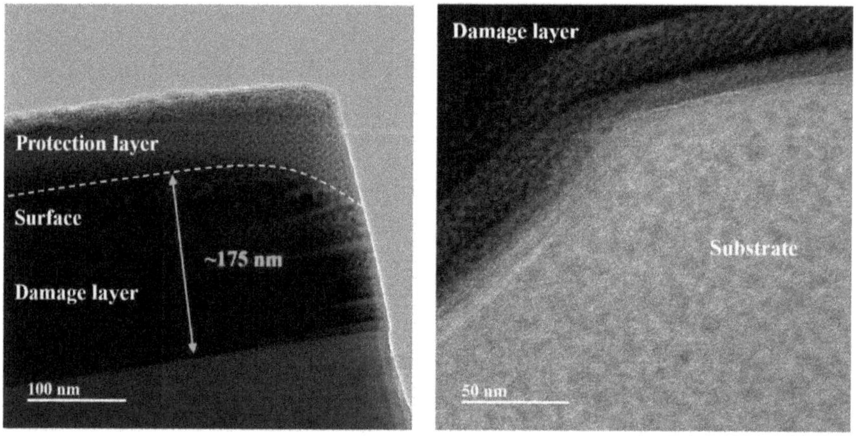

(a) (b)

Figure 12. Subsurface damage of glass-ceramics under 28 W laser-assisted nano-cutting: (**a**) near surface region; (**b**) substrate region.

The subsurface damages of glass-ceramics present diverse characteristics after taper cutting with and without laser assistance. In conventional nano-cutting (Figure 11b), the subsurface crack region contains dense microcracks and crack initiation sites, forming a "source-branch-network" structure. As a typical hard and brittle material, glass-ceramics cannot relieve internal stress through plastic deformation during conventional cutting. Stress concentration triggers crack initiation, and cracks propagate randomly along grain boundaries or defects to a depth of approximately 2 μm (consistent with Figure 10a). These network-like cracks penetrate the subsurface, reducing mechanical strength and optical performance, and require extensive polishing to remove defects. This is consistent with the observation of numerous subsurface cracks in 0 W laser-cut samples, reflecting a typical brittle removal feature.

In 28 W laser-assisted nano-cutting (Figure 12b), there are no subsurface cracks in the subsurface region, and a clear boundary exists between the plastic damage layer and the undamaged substrate. Plastic deformation dissipates most mechanical stress, preventing stress transfer to the substrate and thus inhibiting crack initiation and propagation. As a result, the substrate remains intact without structural damage, preserving the material's original mechanical and optical properties. The distinct damage-substrate boundary avoids deep subsurface damage, which is consistent with Figure 10d, verifying effective crack suppression. The effect of lateral crack propagation and the suppression of subsurface cracks shows a clear qualitative change. As the laser power increases, on one hand, subsurface cracks, even in the brittle fracture area, are less likely to exhibit fracture characteristics. Moreover, even if cracks occur, their propagation trend is significantly suppressed.

In traditional diamond cutting, most of the cutting force, or the energy introduced by the tool, is released through brittle fracture. After processing, there is very little evidence of plastic damage or deformation in the subsurface. However, due to the heating effect of the laser, the material's ability for plastic flow and deformation is significantly enhanced. At this point, most of the energy introduced by the tool is converted into plastic deformation of the material, leading to the formation of a larger plastic deformation damage layer. This is the main change in the material removal mechanism.

Conventional nanoscale cutting of glass-ceramics induces damage that is dominated by brittle fracture, manifested as an amorphous layer on the near-surface and deep subsurface cracks. This damage markedly degrades surface integrity and substantially raises the cost of subsequent polishing. In contrast, 28-W laser-assisted nanoscale cutting converts the material-removal mechanism to a ductile regime, producing only a thin, removable plastically-deformed layer confined to the near-surface region while suppressing the formation of subsurface cracks. Consequently, surface quality is significantly enhanced and polishing requirements are reduced, rendering this approach a reliable method for high-quality machining of glass-ceramic optical components.

6. Conclusions

This study systematically investigates the dynamic material-removal mechanisms of glass-ceramics under laser-assisted nanocutting conditions by integrating numerical simulations with experimental measurements. The main conclusions are as follows:

(1) Laser-heating simulations reveal that an optimal laser power of 30 W raises the temperature of the glass-ceramic to approximately 1000 K.
(2) Nano-cutting simulations indicate that in-situ laser heating markedly improves cutting performance: the average main cutting force decreases by ~17.9%, and the average normal force drops by ~10.5%.
(3) Taper-cutting experiments show that at a laser power of 28 W, the critical NOSC depth reaches 139.44 nm, representing a 281.84% increase compared with conventional cutting, thereby providing critical guidance for process-parameter optimization.
(4) TEM subsurface-damage analysis demonstrates that laser-assisted cutting effectively suppresses crack propagation. At 10 W, subsurface cracks are significantly reduced; at 28 W, no cracks are observed in the crack-free removal zone, and surface integrity improves with increasing laser power. Although a plastic damage layer of approximately 175 nm remains, it can be readily eliminated by subsequent polishing.

Collectively, these results furnish a theoretical foundation for optimizing LADT parameters, confirming the feasibility of high-quality machining of glass-ceramics by the LADT method.

Author Contributions: Conceptualization, J.L.; Validation, J.L. and F.X.; Resources, F.J.; Data curation, J.L.; Writing—original draft, J.L.; Supervision, F.J.; Funding acquisition, F.X. All authors have read and agreed to the published version of the manuscript.

Funding: This work was supported by the National Key Research and Development Program of China (2023YFB3407200) and the National Natural Science Foundation of China (52475498).

Data Availability Statement: The original contributions presented in this study are included in the article. Further inquiries can be directed to the corresponding author.

Conflicts of Interest: The authors declare no conflict of interest.

References

1. Casasola, R.; Rincón, J.M.; Romero, M. Glass–ceramic glazes for ceramic tiles: A review. *J. Mater. Sci.* **2012**, *47*, 553–582. [CrossRef]
2. Serbena, F.C.; Mathias, I.; Foerster, C.E.; Zanotto, E.D. Crystallization toughening of a model glass-ceramics. *Acta Mater.* **2015**, *86*, 216–228. [CrossRef]
3. Huang, H.; Li, X.; Mu, D.; Lawn, B.R. Science and art of ductile grinding of brittle solids. *Int. J. Mach. Tools Manuf.* **2021**, *161*, 103675. [CrossRef]
4. Zhang, X.; Hu, H.; Wang, X.; Luo, X.; Zhang, G.; Zhao, W.; Wang, X.; Liu, Z.; Xiong, L.; Qi, E.; et al. Challenges and strategies in high-accuracy manufacturing of the world's largest SiC aspheric mirror. *Light Sci. Appl.* **2022**, *11*, 310. [CrossRef]
5. Brinksmeier, E.; Gläbe, R.; Osmer, J. Ultra-Precision Diamond Cutting of Steel Molds. *CIRP Ann.* **2006**, *55*, 551–554. [CrossRef]
6. Hashimoto, T.; Yan, J. Effect of System Dynamics on Surface Topography in Fast Tool Servo-Based Diamond Turning of Microlens Arrays. *Nanomanuf. Metrol.* **2025**, *8*, 8. [CrossRef]
7. Wu, K.; Zhang, J.; Zheng, Z.; Li, Z.; Ding, P.; Liu, J.; Wang, J. Fabrication of Micro/Nanostructured Copper Fibers by Vibration Cutting for Felt-Based Freshwater Purification. *Nanomanuf. Metrol.* **2024**, *7*, 21. [CrossRef]

8. Kupczyk, M.; Jozwiak, K.; Cieszkowski, P.; Libuda, P. Influence of laser heating on adhesion of CVD coatings to cutting edge *Surf. Coat. Technol.* **2007**, *201*, 5153–5156. [CrossRef]
9. Wang, J.P.; Zhang, G.Q.; Chen, N.; Zhou, M.H.; Chen, Y.B. A review of tool wear mechanism and suppression method in diamon turning of ferrous materials. *Int. J. Adv. Manuf. Technol.* **2021**, *113*, 3027–3055. [CrossRef]
10. Zhang, Z.; Yan, J.; Kuriyagawa, T. Study on tool wear characteristics in diamond turning of reaction-bonded silicon carbide. *Int. Adv. Manuf. Technol.* **2011**, *57*, 117–125. [CrossRef]
11. An, Q.L.; Ming, W.; Ming, C. Experimental Investigation on Cutting Characteristics in Nanometric Plunge-Cutting of BK7 an Fused Silica Glasses. *Materials* **2015**, *8*, 1428–1441. [CrossRef]
12. You, K.Y.; Liu, G.Y.; Wang, W.; Fang, F. Laser assisted diamond turning of silicon freeform surface. *J. Mater. Process. Technol.* **202** *322*, 118172. [CrossRef]
13. Chen, X.; Liu, C.; Ke, J.; Zhang, J.; Shu, X.; Xu, J. Subsurface damage and phase transformation in laser-assisted nanometri cutting of single crystal silicon. *Mater. Des.* **2020**, *190*, 108524. [CrossRef]
14. Mohammadi, H.; Patten, J.A. Laser Augmented Diamond Drilling: A New Technique to Drill Hard and Brittle Materials. *Procedi Manuf.* **2016**, *5*, 1337–1347. [CrossRef]
15. Langan, S.M.; Ravindra, D.; Mann, A.B. Mitigation of damage during surface finishing of sapphire using laser-assisted machinin *Precis. Eng.* **2018**, *56*, 1–7. [CrossRef]
16. Ravindra, D.; Ghantasala, M.K.; Patten, J.A. Ductile mode material removal and high pressure phase transformation in silico during micro-laser assisted machining. *Precis. Eng.* **2012**, *36*, 364–367. [CrossRef]
17. You, K.Y.; Fang, F.Z.; Yan, G.P. Surface generation of tungsten carbide in laser-assisted diamond turning. *Int. J. Mach. Tools Manu* **2021**, *168*, 103770. [CrossRef]
18. Milhans, J.; Li, D.S.; Khaleel, M.; Sun, X.; Al-Haik, M.S.; Harris, A.; Garmestani, H. Mechanical properties of solid oxide fuel cel glass-ceramic seal at high temperatures. *J. Power Sources* **2011**, *196*, 5599–5603. [CrossRef]
19. Yilbas, B.S.; Akhtar, S.S. Laser Cutting of Alloy Steel: Three-Dimensional Modeling of Temperature and Stress Fields. *Mate Manuf. Process.* **2011**, *26*, 104–112. [CrossRef]
20. Kurgin, S.; Dasch, J.M.; Simon, D.L.; Barber, G.C.; Zou, Q. Evaluation of the convective heat transfer coefficient for minimum quantity lubrication (MQL). *Ind. Lubr. Tribol.* **2012**, *64*, 376–386. [CrossRef]
21. Cui, R.; Zhou, L.; Ren, Q.; Li, L.; Li, W. Optimizing porous glass-ceramics fabricated from coal gasification fine slag: Effects of ash composition on structure and properties. *Constr. Build. Mater.* **2025**, *470*, 140560. [CrossRef]
22. Fedoseev, D.V.; Vnukov, S.P.; Bukhovets, V.L.; Anikin, B.A. Surface graphitization of diamond at high temperatures. *Surf. Coat Technol.* **1986**, *28*, 207–214. [CrossRef]

micromachines

MDPI

rticle

:nhancing the Machinability of Sapphire via Ion Implantation ınd Laser-Assisted Diamond Machining

nyang Ke [1], Honglei Mo [2], Ke Ling [1], Jianning Chu [3], Xiao Chen [3,*] and Jianfeng Xu [1,*]

[1] State Key Laboratory of Intelligent Manufacturing Equipment and Technology, School of Mechanical Science and Engineering, Huazhong University of Science and Technology, Wuhan 430074, China; jinyang_ke@hust.edu.cn (J.K.)

[2] Shanghai Aerospace Control Technology Institute, Shanghai 201109, China

[3] Hubei Key Laboratory of Modern Manufacturing Quality Engineering, School of Mechanical Engineering, Hubei University of Technology, Wuhan 430068, China

* Correspondence: chenxiao1987jz@163.com (X.C.); jfxu@hust.edu.cn (J.X.)

Abstract: Sapphire crystals, owing to their outstanding mechanical and optical properties, which are widely used in advanced optics, microelectronic devices, and medical instruments. The manufacturing precision of sapphire optical components critically affects the performance of advanced optical systems. However, the extremely high hardness and low fracture toughness of sapphire make it a typical hard-to-machine material, prone to brittle surface fractures and subsurface damage during material removal. Improving the machinability of sapphire remains a pressing challenge in advanced manufacturing. In this study, surface modification and enhanced ductility of C-plane sapphire were achieved via ion implantation, and the machinability of the modified sapphire was further improved through laser-assisted diamond machining (LADM). Monte Carlo simulations were employed to investigate the interaction mechanisms between incident ions and the target material. Based on the simulation results, phosphorus ion implantation experiments were conducted, and transmission electron microscopy observation was used to characterize the microstructural evolution of the modified layer, while the optical properties of the samples before and after modification were analyzed. Finally, groove cutting experiments verified the enhancement in ductile machinability of the modified sapphire under LADM. At a laser power of 16 W, the ductile–brittle transition depth of the modified sapphire increased to 450.67 nm, representing a 51.57% improvement over conventional cutting. The findings of this study provide valuable insights for improving the ductile machining performance of hard and brittle materials.

Keywords: sapphire; ion implantation; ultra-precision cutting; laser-assisted diamond machining

1. Introduction

Sapphire possesses outstanding properties such as high strength, thermal stability, excellent transparency, and superior wear and impact resistance [1–3], making it an ideal substrate material for advanced optical systems in the fields of microelectronic devices and medical instruments [4–6]. However, the extremely high hardness and low fracture toughness of sapphire make it a typical hard-to-machine material with pronounced anisotropy [7,8]. Material removal during nanocutting of sapphire often leads to surface fracture and subsurface damage [9,10]. The ductile–brittle transition depth (DBTD) of sapphire generally falls within the range of 100–400 nm, with specific values strongly dependent on crystal orientation [7,11]. Such a small DBTD severely restricts the material removal rate and imposes stringent requirements on machine tool control, making it extremely challenging to achieve a consistent ductile regime cutting across the entire workpiece surface.

Single-point diamond turning (SPDT) is an important technology in ultra-precision manufacturing for fabricating complex optical surfaces. It offers advantages of high precision, high efficiency, and excellent machining flexibility, and has attracted widespread attention over the past few decades [12,13]. Nevertheless, the extreme hardness of sapphire often causes severe tool wear, leading to a dramatic reduction in tool life [11,14,15]. In addition, subsurface damage introduced during machining can deteriorate optical path accuracy and compromise the service performance of optical systems [16]. Currently, the traditional processing route for sapphire optical components involves grinding, lapping and polishing. However, strong grinding forces result in poor surface quality, and the inevitable subsurface damage layer requires complex post-treatment [10,17,18]. Moreover, tool–workpiece interference and edge effects during grinding and polishing hinder the achievement of high formation accuracy [17]. These limitations hinder conventional processes from achieving high-precision and low-damage manufacturing of hard and brittle optical components. They also underscore the urgent need for technological advances beyond traditional methods.

Ion implantation offers a promising strategy for tailoring the mechanical properties and crystal structure of materials, thereby reducing the hardness and brittleness of sapphire. During ion implantation, interactions between high-energy particles and the target material can trigger a variety of important physical processes, including sputtering, backscattering, and the formation of vacancies and interstitial atoms, as illustrated in Figure 1a. Irradiation of sapphire by energetic ions triggers complex collision events. High-energy ions exceeding the displacement threshold displace lattice atoms, generating vacancies, initiating cascades, or causing sputtering. Low-energy ions below the threshold mainly induce lattice vibrations and dissipate energy thermally without permanent damage. In some cases, incident ions may channel through the crystal, forming isolated vacancies at the ends of their trajectories [19,20]. Skuratov et al. [21] investigated four crystallographic orientations of sapphire crystals irradiated with Kr (305 MeV), Xe (595 MeV), and Bi ions (710, 557, 269, and 151 MeV). They reported that obvious surface defects could only be observed under Bi ion irradiation with energies higher than 269 MeV. Kabir et al. [22] conducted irradiation experiments on sapphire crystals using 0.7 MeV Xe ions at doses ranging from 5×10^{11} to 2×10^{14} ions/cm^2. Their results showed that when the dose exceeded 1.2×10^{13} ions/cm^2, the crystal structure became completely disordered starting from the surface. Current research on sapphire modification has mainly focused on defect formation mechanisms, while systematic investigations into the influence of implantation parameters on ion–target interactions remain limited.

High-energy ion implantation can generate various damage structures, ranging from isolated point defects and defect clusters to the formation of amorphous regions [23]. Such damage can reduce the hardness and brittleness of crystals, thereby enabling enhanced plastic deformation during material removal [24]. Fang et al. [19] proposed a novel method termed "nanometric machining of ion-implanted materials" in which fluorine ions were implanted into single-crystal silicon to form a 4.5 μm thick amorphous modified layer, followed by diamond cutting for ductile regime material removal. Their results showed that the DBTD increased to 923.566 nm, the surface roughness Ra reached 0.861 nm, and the tool life was extended by five times. Wang et al. [25] investigated the modification effects of oxygen ion implantation on the nanocutting of single-crystal silicon. Their experimental results showed that the irradiation-induced damage absorbed shear strain energy during the nanocutting process, thereby promoting plastic deformation of the material. To et al. [26] modeled and visualized the distribution of implanted hydrogen ions and the displacement damage induced in the subsurface of single-crystal silicon, and further validated the enhanced machinability of surface-modified silicon through ultra-precision micro-cutting experiments. However, the hardness of silicon is much lower than that of sapphire, highlighting the urgent need to further explore the potential of ion implantation surface modification in improving ductile regime machining of hard and brittle materials.

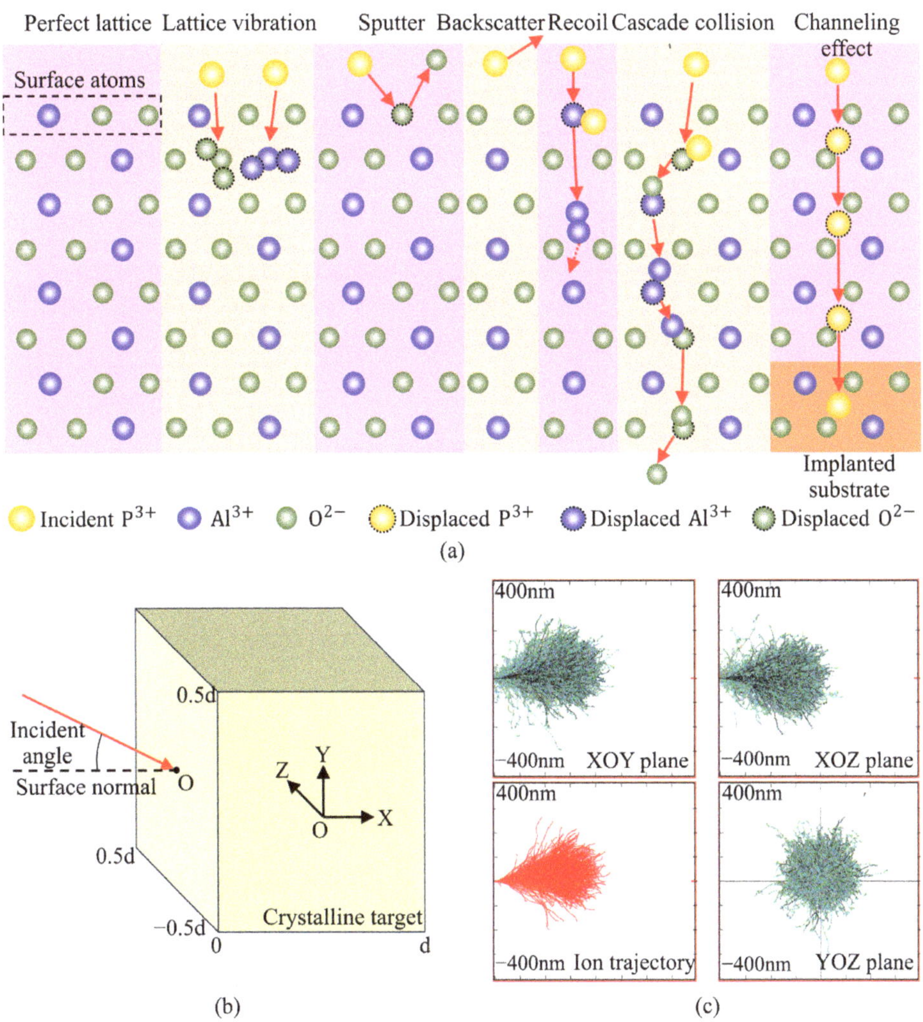

Figure 1. Simulation of energetic ion implantation in sapphire crystal. (**a**) Schematic illustration of the ion implantation process where the red arrows indicate the direction of atomic/ionic motion; (**b**) three-dimensional model used for TRIM simulation; (**c**) snapshot of P ions with energy of 400 keV at an incidence angle of 7°.

Over the past decades, laser-assisted machining (LAM) has demonstrated great potential in processing hard and brittle materials. For metals such as titanium alloys and superalloys, LAM has been shown to effectively reduce yield strength, thereby lowering cutting forces [27], improving surface integrity [28], and extending tool life [29]. Mohammadi et al. [30] proposed laser-assisted diamond machining (LADM), wherein a focused laser beam is transmitted through a transparent tool to irradiate the cutting zone directly, enabling localized and precise heating to enhance the machinability of hard and brittle materials. Numerous experimental studies have verified the effectiveness of LADM for the ultra-precision cutting of silicon [31], silicon carbide [32], and tungsten carbide [33,34]. Reported benefits include reduced cutting forces and tool wear, suppression of surface and subsurface defects, improved machining quality, and enhanced processing efficiency [32,34–36]. However, given the high transmittance of sapphire over a broad spectral range, the effectiveness of LADM of sapphire remains uncertain.

To address the poor machinability of sapphire in ultra-precision cutting, this study employed ion implantation to achieve surface modification and enhance the ductility of C-plane sapphire, followed by the application of LADM to further improve ductile machinability. Monte Carlo simulations were conducted to investigate the ion–target interaction mechanisms during implantation. Based on the simulation results, phosphorus ion implantation experiments were performed, and transmission electron microscopy (TEM) was employed to characterize the microstructural evolution of the ion-induced modified layer. The optical properties of sapphire samples before and after implantation were also analyzed. Finally, groove cutting experiments were conducted to validate the enhancement of the machinability of the modified sapphire under LADM.

2. Materials and Methods

2.1. Basic Principles and Simulation of Ion Implantation

The ion–target interaction process is accompanied by complex energy exchange mechanisms. The displacement energy, E_{disp}, refers to the minimum energy required to displace a target atom sufficiently far from its lattice site so that it cannot quickly return. The lattice binding energy, E_{latt}, is defined as the minimum energy required to remove an atom completely from the lattice. The surface binding energy, E_{surf}, refers to the energy needed to remove an atom from the surface of the target material. Atoms with a final energy, E_{final}, below the threshold are considered stationary within the crystal target. When a moving atom collides with a target atom and transfers energy exceeding E_{disp}, the target atom is displaced from its lattice site. As it loses E_{latt} to the lattice, its recoil energy is given by $E_{recoil} = E_{disp} - E_{latt}$. If the recoiled target atom possesses energy greater than E_{disp}, it can continue to collide with other target atoms, generating additional vacancies, i.e., a collision cascade. For sapphire crystal targets, the default values for aluminum atoms are: $E_{disp} = 25$ eV, $E_{latt} = 3$ eV, and $E_{surf} = 3.36$ eV; for oxygen atoms: $E_{disp} = 28$ eV, $E_{latt} = 3$ eV, and $E_{surf} = 2$ eV [37,38].

In this study, the TRIM (Transport of Ions in Matter) Monte Carlo simulator from the SRIM-2013 package, developed by J.F. Ziegler and J.P. Biersack, was employed to calculate the collision processes between high-energy ions and the target material [39]. The simulator describes ion–target collisions using the Ziegler–Biersack–Littmark (ZBL) universal potential [40] and establishes a complete theoretical framework based on extensive experimental data. Figure 1b shows the three-dimensional model and coordinate system used for the TRIM simulations. The target material was modeled as a cube with a length of $d = 800$ nm, and energetic phosphorus (P) ions were obliquely incident on the target surface with energies ranging from 100 to 600 keV. TRIM features a graphical user interface and a comprehensive compound target database, allowing the simulation of various physical processes, including target damage, ion ranges, backscattering yields, sputtering yields, ionization, and phonon generation [41]. Each simulation tracked 1000 ions to ensure statistical reliability. The "detailed full-cascade" mode was employed to simulate the ion implantation experiments, providing data on backscattering yields, vacancy numbers, ion range distributions, and energy exchange processes. All ion trajectories were traced using the binary collision approximation. Figure 1c presents a snapshot of P ions with an energy of 400 keV at an incidence angle of 7° at the end of the implantation process. The primary source of energy loss arises from nuclear stopping, while incident ions continuously and uniformly lose energy through interactions with target electrons in the intervals between collisions.

2.2. Subsurface Structural Characterization

The experimental specimens were procured from TDG Holding Co., Ltd., (Haining, China), with a diameter of 34 mm and a thickness of 3 mm. It should be noted that the selected implantation dose in this study ensures sufficient amorphization of the substrate over a larger depth range, as theoretically calculated in our previously published study [42]. TEM samples were extracted from specific regions of grooves obtained in groove cutting experiments using a focused ion beam (FIB) system (Helios Nanolab 660, FEI Company, Hillsboro, OR, USA). Prior to preparation, the workpieces were mounted on conductive tape and sputter-coated with gold for 240 s to enhance conductivity, followed by deposition of a tungsten protective layer to prevent potential damage during FIB processing. Aberration-corrected TEM (Titan G2 60–300 probe, FEI Company, Hillsboro, OR, USA) was employed to examine the sample cross-sections and perform selected-area electron diffraction (SAED) analysis, enabling characterization of subsurface microstructural changes induced by ion implantation.

2.3. Groove Cutting Tests

In LADM, a focused laser beam passes through a transparent tool and directly irradiates the cutting edge to locally and precisely heat the material, as shown in Figure 2a. This localized laser irradiation reduces the cutting resistance of brittle and hard-to-machine materials through thermal softening, promotes shear deformation, and facilitates chip formation along the rake face. The light source in LADM is a continuous 1070 nm fiber laser. In the experiments, the laser spot was precisely aligned with the tool tip, and the output power at the cutting edge was measured using a power meter (30 (150) A-BB-18, Ophir, Jerusalem, Israel). A custom in situ laser-assisted diamond tool with a −35° rake angle and a 0.506 mm tool tip radius was employed for constant-speed lateral feed. During machining, the cutting depth of each groove was continuously increased from zero to the maximum, as illustrated in Figure 2b. Based on the maximum X-axis travel and the dynamic response of the feed system, a cutting speed of 200 mm/min was chosen to ensure optimal machine performance. Each experiment was repeated three times to ensure data reliability. In the self-developed LADM setup, the maximum stable output power measured at the diamond tool cutting edge was 20 W. To prevent high-temperature failure of the laser and ensure stable operation of the cutting system, the maximum power applied in the groove cutting experiments was set to 16 W. The parameters of the groove cutting experiments are listed in Table 1. It is worth noting that in this study, the cutting speed was kept constant while the laser power was set at five evenly spaced levels, in order to intuitively demonstrate the enhancement of ductile cutting performance of modified sapphire under LADM.

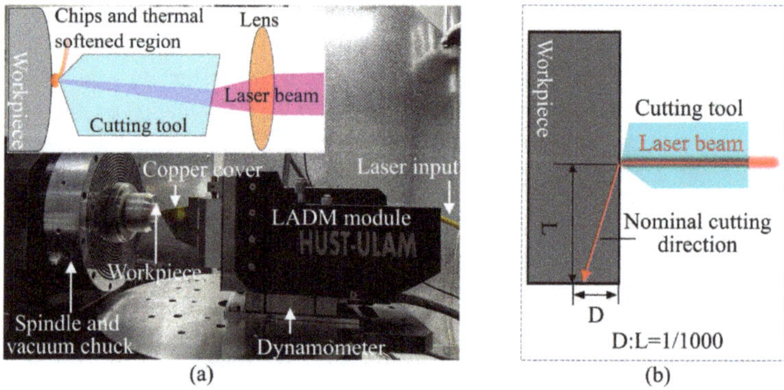

Figure 2. Experimental validation of LADM cutting. (**a**) Photograph of the experimental setup; (**b**) schematic illustration of the groove cutting experiment.

Table 1. Parameters of groove cutting experiments.

Description	Value
Workpiece	C plane sapphire
Laser source	1070 nm continuous-wave laser
Tool nose radius	0.506 mm
Rake angle	−35°
Flank angle	10°
Cutting speed	200 mm/min
Groove slope	D/L = 1:1000
Laser power	0 W, 4 W, 8 W, 12 W, 16 W

2.4. Characterization Method

The grooves obtained from groove cutting experiments with varying cutting depths were observed using an optical microscope (DSX 510, Olympus Corporation, Tokyo, Japan) to examine their surface morphology. The DBTD of the grooves was measured using a white-light interferometer (NewView 9000, ZYGO Corporation, Middlefield, CT, USA). DBTD was defined as the depth at which the first crack appeared on the machined surface. The chips generated during machining were collected and examined using a scanning electron microscope (GeminiSEM 360, ZEISS Group, Oberkochen, Baden-Württemberg, Germany).

3. Results and Discussions

3.1. Effect of Energy on Ion–Target Interactions

Figure 3 illustrates the effect of the incidence angle on the implantation behavior of P ions in sapphire crystals. First, as the incident energy increases from 100 keV to 500 keV, the backscattering fraction gradually increases with increasing incidence angle. In contrast, the vacancy-to-ion ratio exhibits the opposite trend, decreasing progressively with increasing incidence angle. Notably, when the incidence angle is less than 40°, both parameters remain relatively stable. The backscattering fraction stays near zero, while the vacancy-to-ion ratio reaches about 1000 at 100 keV and 2700 at 500 keV. When the incidence angle exceeds 40°, the backscattering fraction increases exponentially with the angle. This observation indicates that ion implantation is more stable under small-angle incidence conditions. In engineering practice, the incidence angle is typically set to 7° to avoid channeling effects and ensure uniform implantation. Accordingly, a fixed incidence angle of 7° was used in both the simulations and experiments of this study. This approach not only minimizes ion scattering losses but also ensures a sufficient defect generation rate within the modified layer.

Figure 4 shows the depth profiles of P ions implanted at different incident energies under a fixed incidence angle of 7°. It can be observed that, as the incident energy increases, the peak positions of the P ion distributions gradually shift deeper into the target, indicating that the average penetration depth of the ions increases with energy. At the same time, the shape of the ion distribution curves changes. The peak concentration decreases, while the distribution broadens. This indicates that high-energy ions undergo a longer energy dissipation process within the target, resulting in a more dispersed spatial distribution of implanted ions. This behavior reflects a fundamental phenomenon in ion implantation: under the same material and incidence angle conditions, the projected range of ions increases nonlinearly with incident energy, while the local peak concentration is diluted due to the extended penetration depth. It is noteworthy that the unit of ion concentration is expressed as (Atoms/cm^3)/(Atoms/cm^2), representing the probability density of ion distribution per unit implantation dose. Although this unit does not directly correspond to the absolute concentration, multiplying it by the actual implantation dose (in ions/cm^2) yields the true dopant concentration (in Atoms/cm^3). Therefore, these results provide a theoretical basis for selecting appropriate implantation energy and dose combinations to achieve a desired dopant depth and concentration.

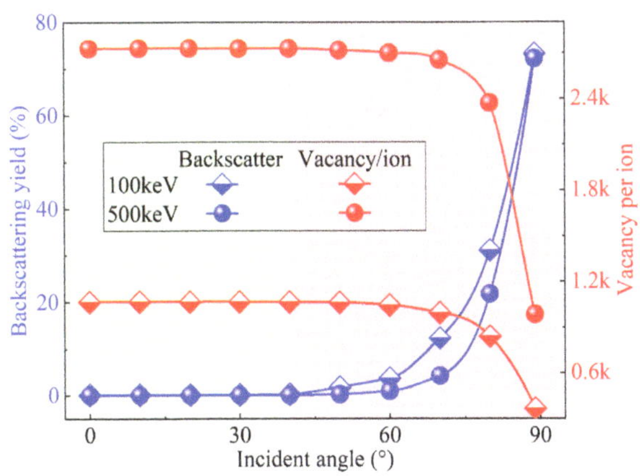

Figure 3. Effect of incident angle on implantation behavior at different incident energies.

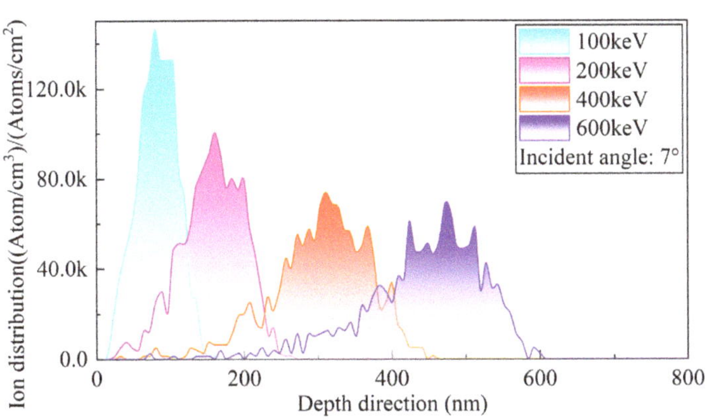

Figure 4. Depth distributions of P ions at different incident energies with an incidence angle of 7°.

To investigate the three-dimensional migration behavior of ions, Figure 5 analyzes the projected ranges of P ions in the three orthogonal planes (XOY, XOZ, and YOZ), corresponding to the depths at which peak concentrations occur. It is evident that, as the ion energy increases, the projected ranges in all three directions increase significantly. This is because high-energy ions travel longer paths within the target, undergo more collisions, and thus exhibit a broader spatial distribution. Comparing different directions, the longitudinal projection (within the XOY plane) remains significantly greater than the transverse (XOZ) and radial (YOZ) projections. This difference arises from the fact that ions predominantly propagate along the incidence direction with limited scattering angles, resulting in a penetration depth along the normal that is much greater than the lateral spread within the plane. In addition, the 7° oblique incidence induces a slight lateral shift, causing a certain asymmetry in the ion distribution within the transverse plane, as shown in Figure 1c.

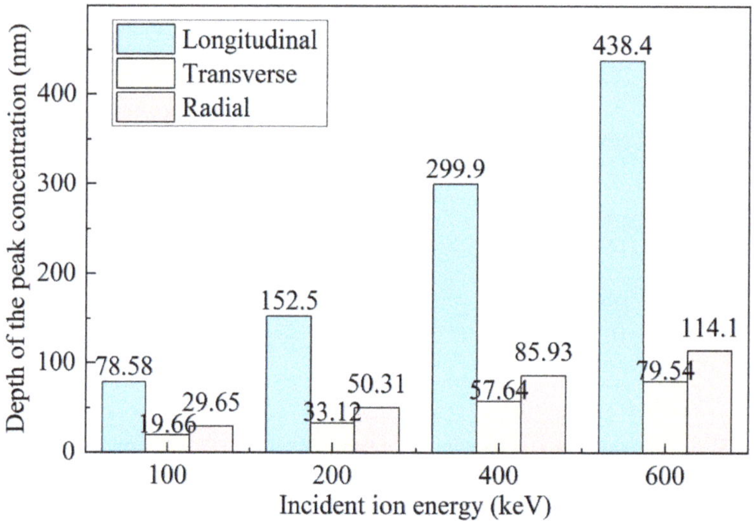

Figure 5. Statistical analysis of the peak concentration depths of P ions.

The regulation of recoil atom behavior by the incident ion energy is a key physical mechanism for achieving ordered microstructural evolution and constructing an ideal modified state in materials. Figure 6 illustrates the distributions of recoiled target atoms induced by P ions at different incident energies. As the ion energy increases, the overall density of recoiled atoms decreases slightly. However, their extension along the depth direction becomes more pronounced, resulting in a marked broadening of the spatial distribution. This trend indicates that high-energy ions can trigger deeper atomic collision events, causing more profound perturbations to the underlying crystal structure and potentially altering the macroscopic mechanical properties of the material. Regarding the composition of recoiled atoms, the density of oxygen recoils is consistently higher than that of aluminum. This can be explained from the perspectives of crystal structure and atomic mass. On one hand, oxygen atoms constitute a higher proportion of the sapphire lattice, making them more likely to undergo recoil events during collisions. On the other hand, because oxygen atoms are lighter and have a lattice binding energy comparable to aluminum, they are more easily displaced by ion collisions. In contrast, the heavier aluminum atoms tend to remain stable.

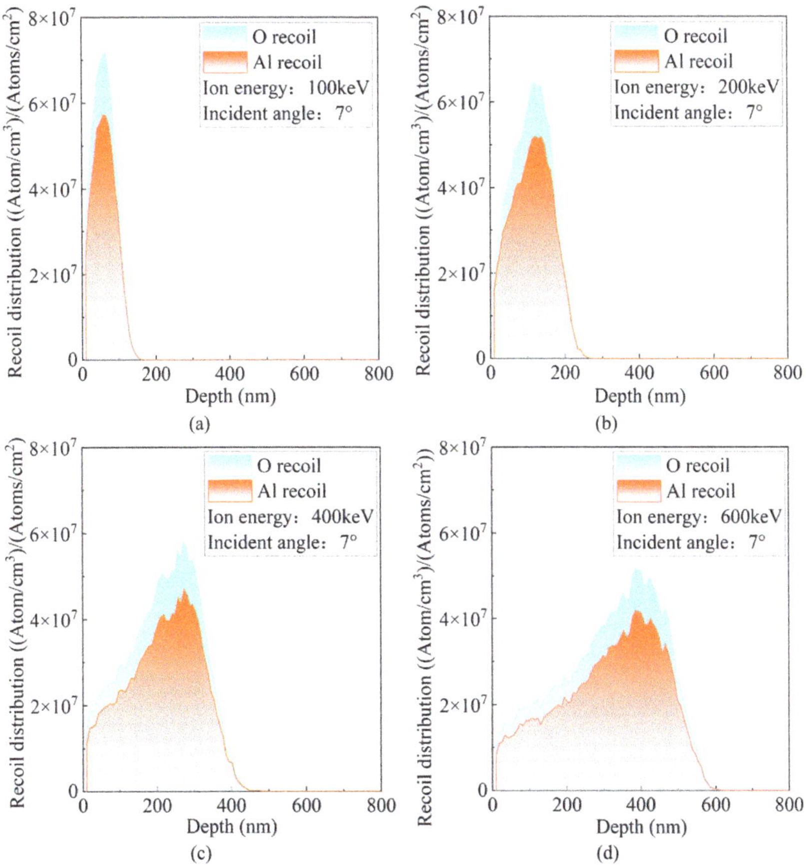

Figure 6. Recoil distributions of target atoms at different incident energies. (**a**) 100 keV; (**b**) 200 keV; (**c**) 400 keV; (**d**) 600 keV.

When ions or recoiled atoms with different kinetic energies collide within a crystal, their effects on target atoms depend on whether the transferred energy exceeds the displacement energy E_{disp}. If the energy transferred during a collision is below this threshold, the target atom may undergo slight displacement or vibration, but not enough to leave its original lattice site. Such collisions do not create permanent defects; instead, they induce local lattice perturbations, manifesting as short-lived oscillations of atoms around their equilibrium positions. In this process, the kinetic energy is no longer propagated as momentum but is converted into phonons via lattice vibrations. Due to strong coupling between atoms in the crystal, the vibration of one atom induces synchronous vibrations in neighboring atoms, generating phonons that propagate as waves. When an atom is displaced from its original lattice site, its binding energy is transferred to phonons generated by the recoiled atom. The remaining phonons are produced by ions or recoiled atoms that transfer energies below E_{disp} to lattice atoms. As shown in Figure 7, at low ion energies, recoiled atoms are mainly concentrated near the surface, and the corresponding phonon energy deposition is similarly localized. With increasing ion energy, recoiled atoms can penetrate deeper into the lattice, triggering collisions and causing the spatial distribution of phonons to extend further into the material. The peak depth of phonon energy deposition shifts deeper with increasing ion energy, indicating that the region of active atomic vibrations within the material also moves inward. This is because high-energy ions are more likely to induce oscillations of deeper atoms. Simultaneously, the peak intensity of phonon energy deposition slightly decreases, suggesting that the number of locally excited phonons per unit volume is reduced. In other words, energy is distributed more uniformly, and local accumulation effects are diminished. Further analysis reveals that incident ions themselves contribute negligibly to phonon generation, which is almost entirely produced by recoiled target atoms.

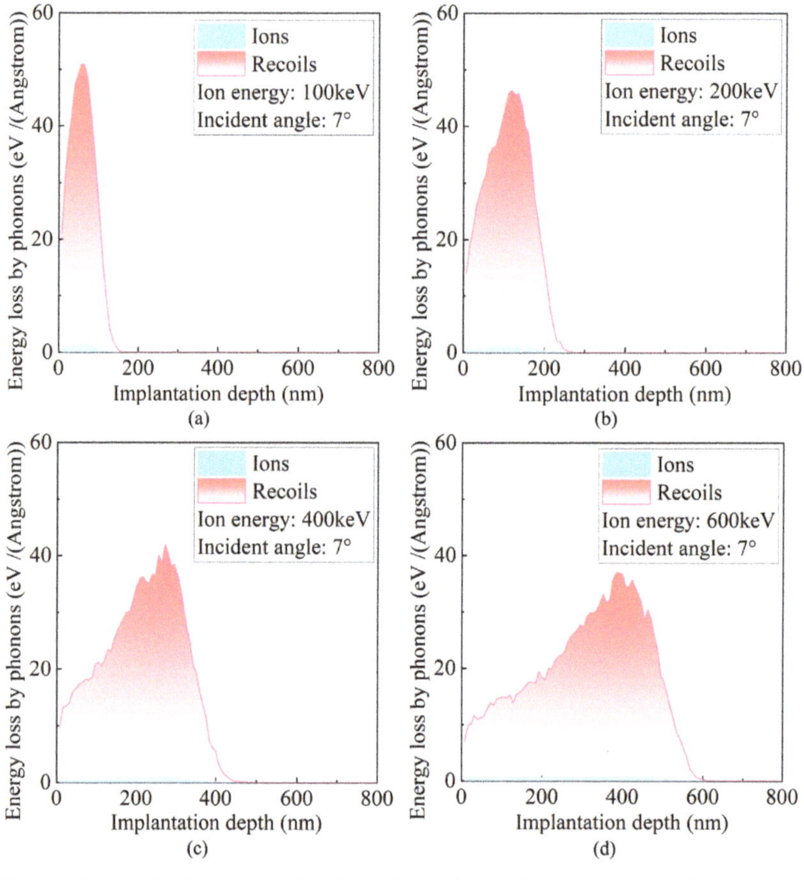

Figure 7. Energy distributions transferred via phonons by incident ions and recoiled target atoms at different incident energies. (**a**) 100 keV; (**b**) 200 keV; (**c**) 400 keV; (**d**) 600 keV.

Figure 8 presents the energy distribution transferred via ionization from incident ions and recoiled target atoms at different incident energies. Ionization energy loss refers to the energy dissipated by ions due to interactions with the target's valence electrons during their motion, representing another major mechanism of energy dissipation in ion–material interactions. As P ions penetrate the crystal and interact with valence electrons, the extent of energy loss depends on the stopping power of these electrons. The simulation results show that, with increasing incident energy, the ionization energy losses induced by both incident ions and recoiled target atoms increase. This trend reflects that high-energy particles possess longer trajectories and higher kinetic energies within the crystal, enabling more inelastic collisions with electrons and resulting in more extensive and intense electronic excitation. Moreover, the ionization energy loss of incident ions is consistently higher than that of recoiled atoms, primarily due to the velocity difference between the two. Although both phonon and ionization energy dissipation mechanisms do not directly generate structural defects, their excitation intensity and spatial distribution significantly influence the thermal response, stress relaxation mechanisms, and local temperature rise during ion implantation. These effects provide important guidance for the optimization of ion implantation process parameters.

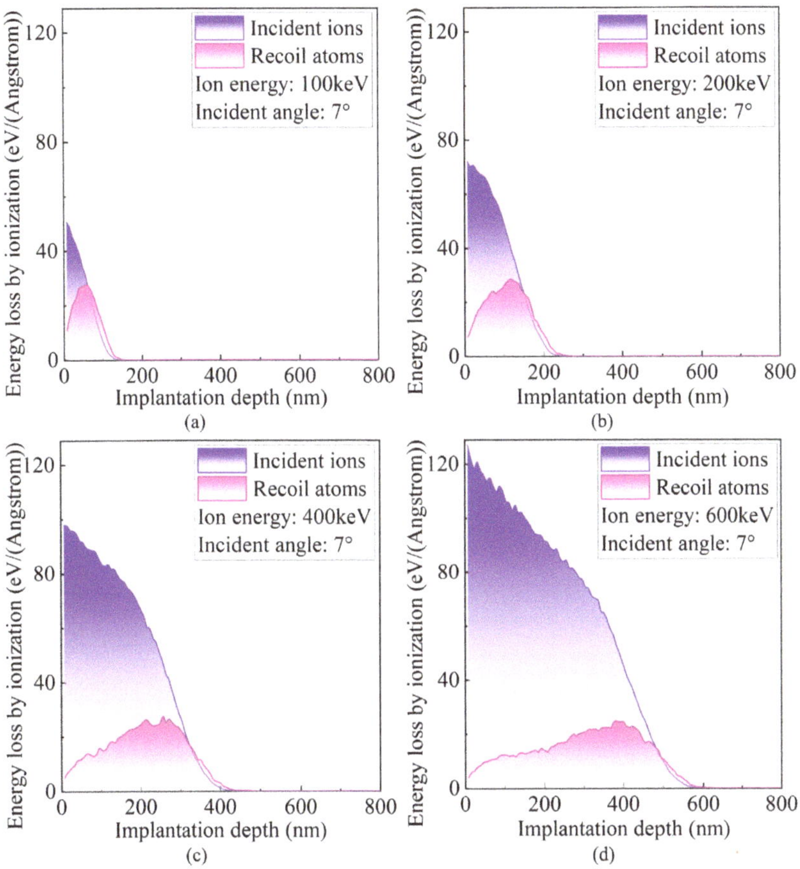

Figure 8. Energy distributions transferred via ionization by incident ions and recoiled target atoms at different incident energies. (**a**) 100 keV; (**b**) 200 keV; (**c**) 400 keV; (**d**) 600 keV.

Figure 9 illustrates the distributions of total deposited ion energy and energy absorbed by target atoms at different incident energies. As the incident energy increases from 100 keV to 600 keV, the peak depth of energy deposition shifts significantly deeper into the material, and the distribution range broadens. This trend indicates that high-energy ions possess greater penetration capability and a wider energy release zone, with energy no longer confined to the near-surface region but gradually diffusing deeper into the material, thereby affecting a larger volume of the crystal structure. Within the main region of energy deposition, corresponding to the average ion range, the energy absorbed by target atoms changes relatively smoothly, suggesting that incident ions continuously transfer energy to lattice atoms during deceleration. Near the stopping end of the ions, the kinetic energy is rapidly exhausted, resulting in a sharp decrease in transferred energy. Under all implantation conditions, the energies absorbed by oxygen and aluminum atoms are comparable, showing no significant bias. To evaluate the utilization efficiency of deposited ion energy, the ratio of the total energy absorbed by oxygen and aluminum atoms to the total deposited ion energy was quantified. The results indicate that this ratio approaches 100% for all incident energies, meaning that nearly all deposited energy is ultimately absorbed by target atoms. From the perspective of material modification, this highly localized and efficient energy absorption process facilitates precise control over the modified layer structure, enabling manipulation of surface stress states, induction of crystal phase transformations, or suppression of crack propagation, and thus serves as a critical mechanism for achieving controllable microstructural evolution.

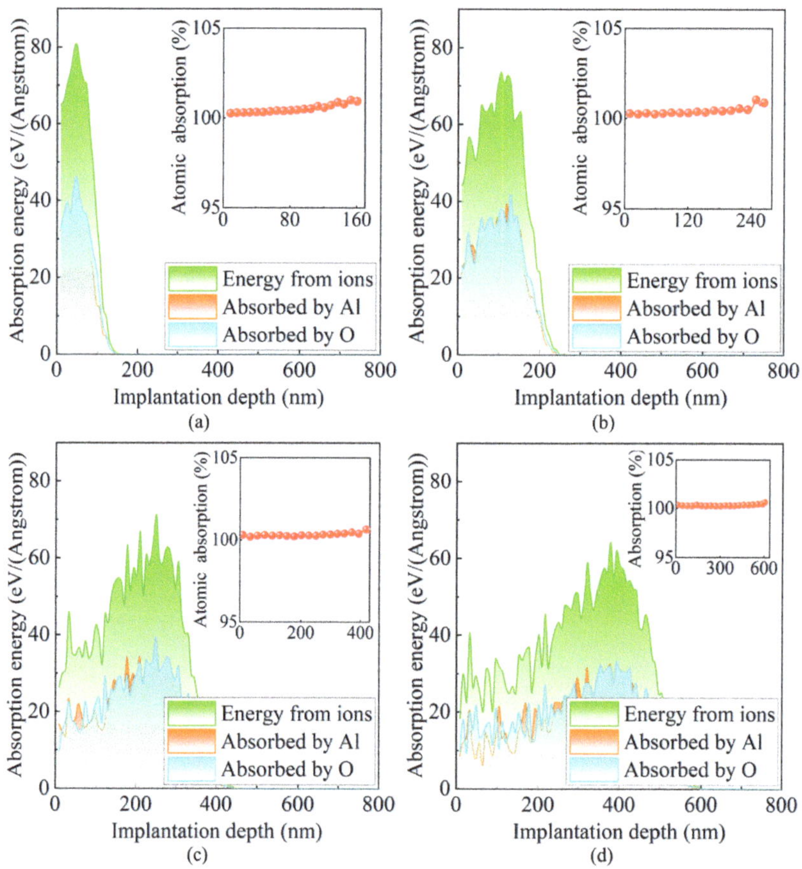

Figure 9. Distributions of deposited ion energy and energy absorbed by target atoms at different incident energies. (**a**) 100 keV; (**b**) 200 keV; (**c**) 400 keV; (**d**) 600 keV. The red dashed curve in the inset shows the ratio of the total energy absorbed by oxygen and aluminum atoms to the total deposited ion energy.

Based on the energy distributions of ions and recoiled atoms described above, the energy dissipation pathways and proportions during ion implantation were further analyzed, as shown in Figure 10. At different incident energies, at least 53% (up to 78.09%) of the total energy of incident particles is consumed through vacancy formation, thereby inducing defects within the crystal target, while at least 20% (up to 43.78%) is ultimately dissipated as phonons. This indicates that phonon dissipation plays a dominant role in ion implantation and represents the primary mechanism for converting kinetic energy into internal energy. Further analysis shows that, with increasing incident energy, the energy transferred to phonons by recoiled atoms decreases, whereas the energy transferred by incident ions gradually increases. These results demonstrate that varying the incident energy significantly affects the dominant phonon energy transfer mechanism. High-energy phonons generated by atomic collisions lead to local temperature rises, which subsequently diffuse to surrounding lower-temperature regions, enabling outward energy transfer.

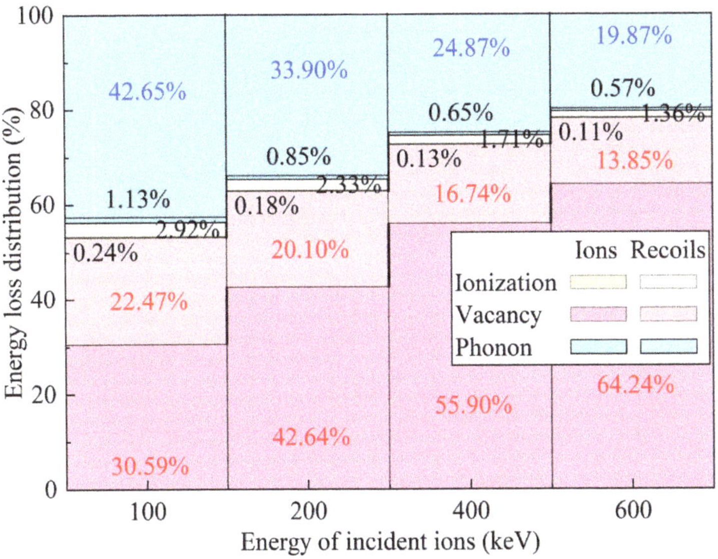

Figure 10. Energy dissipation distributions during the ion implantation process.

Figure 11 shows the spatial distributions of various types of damage, including vacancies, displacements, and replacement collisions, formed in the target material under different incident energies. It can be observed that, as the incident ion energy increases, the spatial extent of all damage types gradually expands, indicating that the damage penetration depth increases with energy. Meanwhile, the peak intensity decreases, suggesting that the damaged energy is spread over a larger volume, reducing local concentration. Notably, the combined intensity of oxygen and aluminum vacancies equals the total vacancy intensity of target atoms. In comparison, the overall vacancy curve lies below the corresponding displacement curve, indicating that the number of vacancies is smaller than the number of displacements, and not all displacement events produce stable vacancies. The bottommost curve represents the distribution of replacement collisions, which occur when incident atoms lose almost all their energy and can no longer advance, filling the sites left by recoiled target atoms. In other words, an incident atom displaces a target atom and then occupies its lattice site. Since they are the same element, no net change occurs in the target. Based on the damage distribution data in Figure 11, the total damage in the target at different incident energies was quantified by integration, as shown in Figure 12. It is noteworthy that, under all conditions, the total number of displacements equals the sum of vacancies and replacement collisions.

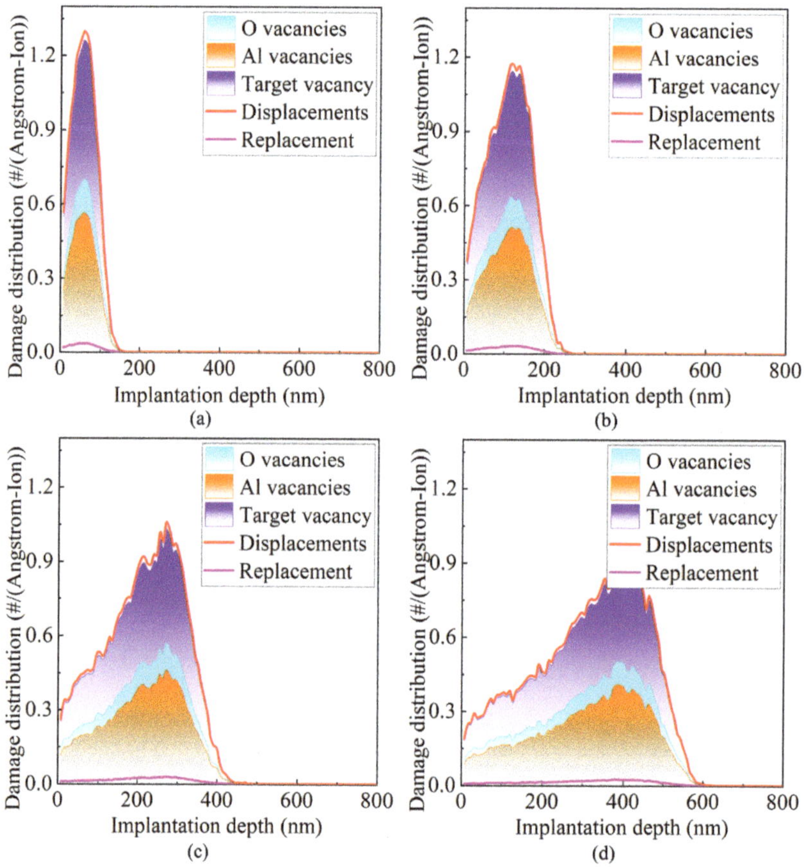

Figure 11. Spatial distributions of various types of damage formed in the target under different incident energies. (**a**) 100 keV; (**b**) 200 keV; (**c**) 400 keV; (**d**) 600 keV.

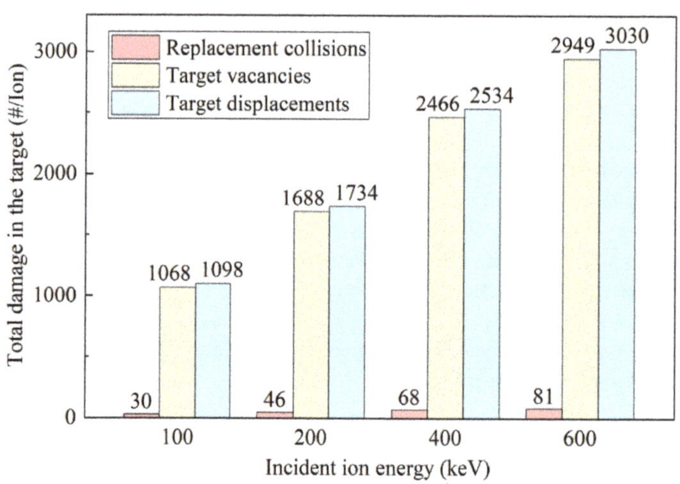

Figure 12. Statistical analysis of the total number of displacements, vacancies, and replacement collisions at different incident energies.

3.2. Microstructural Evolution of the Modification Layer

Based on the Monte Carlo simulation results, phosphorus ions were implanted into C-plane single-crystal sapphire at room temperature using a medium-beam ion implanter (CI-S300MS) developed by China Electronics Technology Group Corporation (Beijing, China), with an incidence angle of 7°. To balance time efficiency and equipment performance, an implantation energy of 300 keV was selected. To ensure the formation of a homogeneous and stable modification layer, the ion implantation dose (i.e., number of ions per unit area) must be carefully optimized. Figure 13a illustrates, using different colors, the spatial distribution of recoil cascades and ion trajectories. The incident ions generate continuous lattice damage, while the cascades triggered by recoiled target atoms result in the formation of defect clusters. Figure 13b depicts the vacancy intensity profile in sapphire subjected to 300 keV phosphorus ion implantation. A_{vacancy} denotes the mean number of vacancies created per unit ion range. Multiplying this value by the implantation dose yields the vacancy concentration per unit volume, expressed as follows:

$$D_{vacancy} = 10^8 \times A_{vacancy} \times n_{dose}, \tag{1}$$

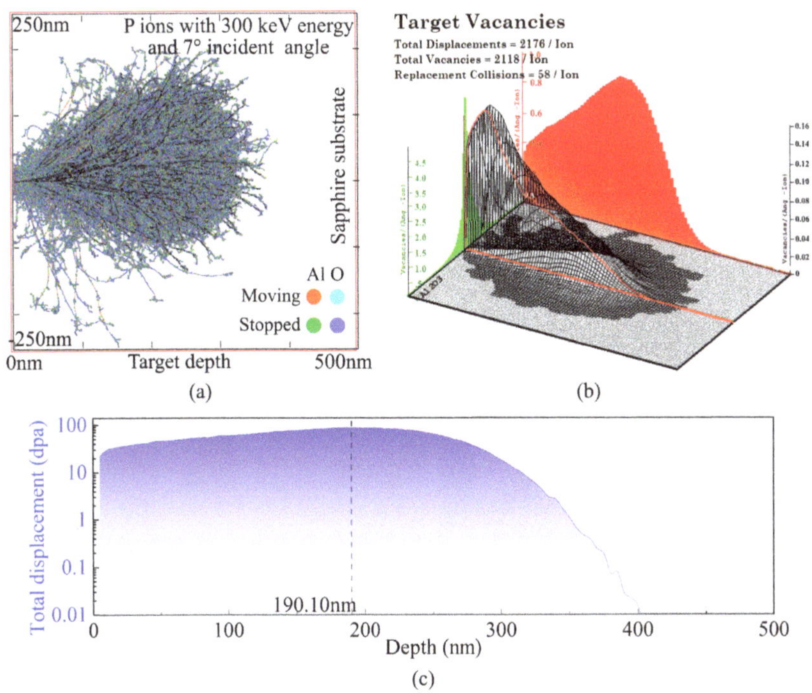

Figure 13. TRIM simulation results of 300 keV phosphorus ion implantation into a sapphire crystal at an incident angle of 7°. (**a**) Visualization of recoil cascade processes and ion trajectories; (**b**) three-dimensional distribution of target vacancies; (**c**) depth profile of total displacement damage.

Here, $A_{vacancy}$ is obtained from TRIM simulations, and n_{dose} represents the implantation dose. Once $D_{vacancy}$ exceeds the atomic density of sapphire (1.175×10^{23} atoms/cm^3), the material undergoes an amorphization transition. As shown in Figure 13b, the peak vacancy intensity along depth reaches ~1.01, implying that amorphization at a depth of 190.10 nm requires a minimum implantation dose of 1.163×10^{15} ions/cm^2. However, since the vacancy intensity near the surface is only ~0.26, a much higher dose would be necessary to achieve full amorphization across the entire 400 nm depth. To guarantee sufficient amorphization of the substrate over a broader depth range, this study adopts 1% of the peak vacancy intensity as a reference, ultimately determining the implantation dose to be 1×10^{17} ions/cm^2. Considering both the implantation dose and the atomic density of sapphire, the following equation presents the calculation method for displacement damage ω (displacements per atom, dpa) derived from TRIM output [43]:

$$\omega = A_{vacancy}\frac{10^8 n_{dose}}{\rho}, \quad (2)$$

where ρ is the atomic density of sapphire. Figure 13c shows that the displacement damage extends over a depth of ~400 nm, reaching a maximum of 86.16 dpa at 190.10 nm.

The microstructural evolution mechanism of the modified layer was systematically characterized using TEM observation. Figure 14a illustrates the evolution of the subsurface microstructure. SAED patterns obtained along the [$\bar{2}$110] zone axis beneath the W protective layer (inset of Figure 14a) confirm the presence of crystalline phases. As shown in Figure 14b, a uniform modified layer with a thickness of 461.18 nm is observed beneath the protective layer. It should be noted that the measured modification depth is about 60 nm greater than the simulated result. This discrepancy arises because TRIM does not account for damage accumulation, which lowers the displacement energy. As the lattice becomes partially disordered, subsequent atoms are more easily displaced by incident ions, thereby intensifying the damage. Such dynamic changes in crystal integrity are beyond the capability of static TRIM simulations, resulting in an underestimation of damage formation. The fast Fourier transform (FFT) pattern of the R2 region (Figure 14c) exhibits characteristic diffuse ring features. As the observation area transitions from the shallow layer to the deeper bulk material, the FFT patterns display a regular matrix of diffraction spots (Figure 14d), consistent with the SAED pattern in the inset of Figure 14a, indicating that the portion of the modified layer near the bulk retains a high degree of crystallinity.

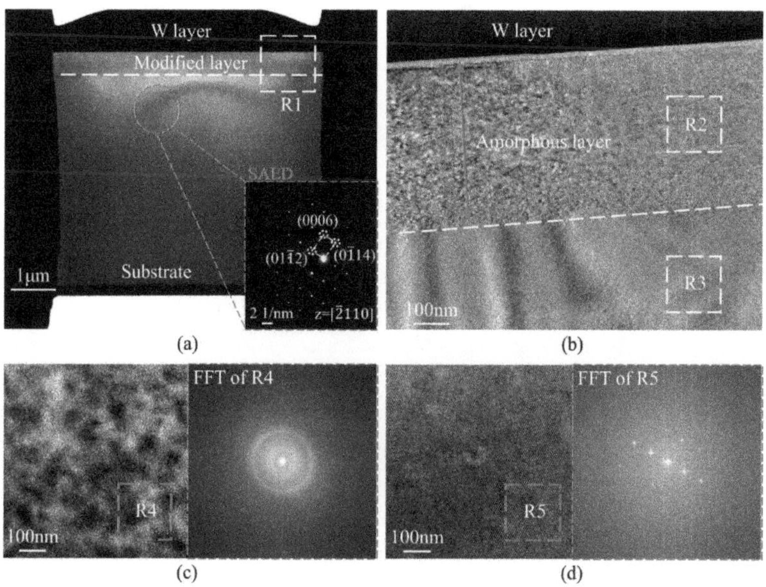

Figure 14. Microstructure of the modified layer in sapphire after phosphorus ion implantation. (**a**) Macroscopic morphology of the TEM sample; (**b**) magnified view of region R1; (**c**) local view of region R2; (**d**) local view of region R3.

The optical properties of sapphire samples before and after surface modification by ion implantation were measured in the near-infrared region (900–1200 nm) using a UV-Vis spectrophotometer (SolidSpec-3700, Shimadzu Corporation, Kyoto, Japan), including transmittance, reflectance, and absorbance, as shown in Figure 15. Taking the wavelength of 1070 nm as an example, the reflectance of the sample surface changed slightly from 14.05% to 14.27% after modification; the transmittance decreased from 85.56% to 80.99%, while the absorbance increased from 0.39% to 4.74%. Clearly, the decrease in transmittance is approximately consistent with the increase in absorbance. This phenomenon indicates that the microscopic defects introduced by ion implantation within the sapphire crystal enhance photon absorption in the near-infrared range, resulting in reduced transmitted light intensity.

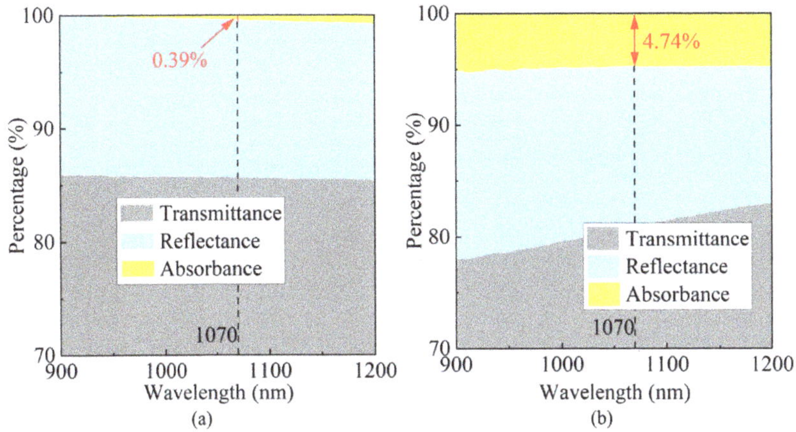

Figure 15. Optical properties of the C-plane sapphire crystal surface in the near-infrared region. (**a**) Before and (**b**) after modification.

3.3. Ductile–Brittle Transition Behavior

Based on the finding that microscopic defects introduced by ion implantation enhance photon absorption in the near-infrared region of sapphire crystals, the effect of laser power on the machinability of modified sapphire was further investigated. Figure 16a presents the surface morphology of sapphire after LADM. Clearly, the micro-grooves generated along the cutting direction exhibit distinct brittle and ductile zones. At the initial stage of the groove, the surface is smooth and clean, with well-defined edges, indicating that the material removal mechanism is predominantly ductile. As the cutting depth increases, pits caused by brittle fracture gradually appear and their density continues to rise, eventually making brittle removal the dominant mechanism. The green solid lines in the figures mark the ductile–brittle transition locations, beyond which the groove surface is covered with cleavage patterns and severe tearing, characteristic of brittle defects as indicated by the yellow arrows. Moreover, with increasing laser power, the extent of the ductile region slightly expands, while the size and density of brittle fracture defects on the groove surface decrease marginally at higher powers. This effect can be attributed to the enhanced thermal softening at higher laser power, which improves the material's plastic deformation capability, allowing chip formation to occur more easily through plastic flow.

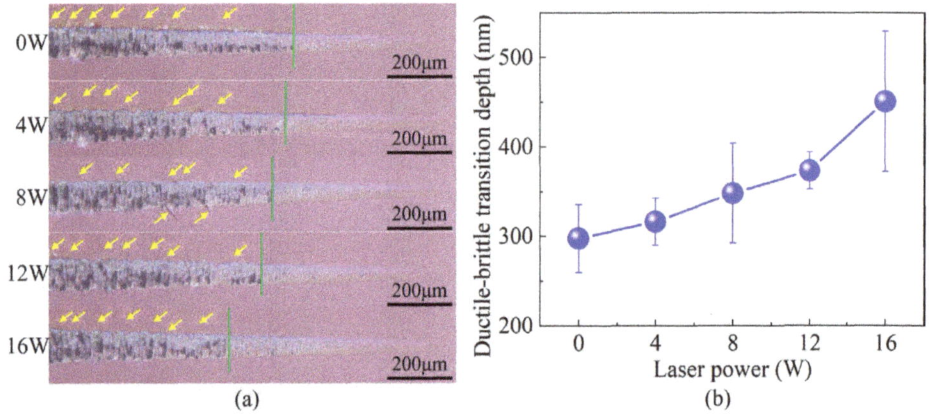

Figure 16. Analysis of surface morphology of modified sapphire after LADM. (**a**) Macroscopic morphology of the grooves where the yellow arrows indicate the locations of surface cracks; (**b**) effect of laser power on the DBTD.

Figure 16b compares the DBTD of modified sapphire under different laser powers. In conventional cutting (without laser assistance), the DBTD of the modified sapphire was 297.33 nm, which is consistent with previously reported experimental results [7,44–46]. With increasing laser power, the DBTD increased, reaching 450.67 nm at 16 W, representing a 51.57% improvement over conventional cutting. It is noteworthy that the DBTD during laser-assisted diamond machining of modified sapphire exhibits a relatively large standard deviation, indicating some instability in the cutting process. This is attributed to the fact that, although the absorption of the modified sapphire at the 1070 nm wavelength used for laser assistance is enhanced, the absolute value remains modest. Future work will focus on process optimization to further improve the machinability of the modified sapphire.

To further investigate the evolution mechanism of material removal modes, Figure 17 shows the morphology of chips generated during LADM of modified sapphire at different laser powers. In conventional cutting (0 W, without in situ laser assistance), the chips primarily exhibit irregular fragments with micron-sized discontinuous block features of random shapes. Some areas display ripple patterns caused by stress waves, reflecting the spalling and fracture of the material during removal, as shown in Figure 17a. These observations indicate that during conventional cutting of implanted sapphire, the removal process is dominated by crack propagation and cleavage fracture. As illustrated in Figure 17b,c, with increasing laser power, the amount of blocky spalling in the chips gradually decreases. At 8 W, a small fraction of ribbon-like chips appears, indicating a transition from blocky fragments to mixed chips. With further power increase, the chip morphology evolves from needle-like to ribbon-like, as shown in Figure 17d,e, with almost no blocky spalling observed. These results further confirm that LADM can enhance the ductility of implanted sapphire.

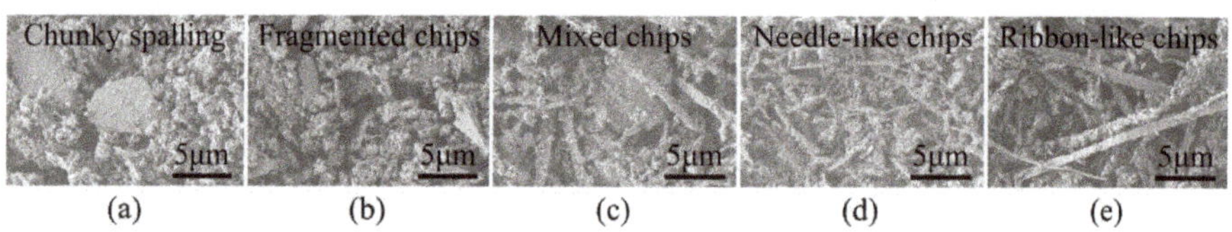

Figure 17. Chip morphology of ion-implanted sapphire after LADM at different laser powers. (**a**) 0 W; (**b**) 4 W; (**c**) 8 W; (**d**) 12 W; (**e**) 16 W.

4. Conclusions

This study investigated the cutting performance of laser-assisted diamond machining (LADM) of ion-implanted sapphire through simulation and experiments. Monte Carlo simulations were employed to examine the effects of implantation parameters, such as incidence angle and ion energy, on ion distribution, damage formation, and energy transfer, thereby elucidating the interaction mechanisms between high-energy ions and the target material. Phosphorus ions were implanted into sapphire crystals, and the microstructural evolution of the resulting modified layer was characterized using transmission electron microscopy (TEM) observation. Finally, groove cutting experiments were conducted to validate the improvement in machinability of the modified sapphire under LADM. The main conclusions drawn from the combined simulation and experimental results are as follows:

(1) The incidence angle and energy selection in the ion implantation process significantly affect the interaction mechanisms between ions and the target atoms. With increasing energy, the projected range of incident ions increases, the distributions of recoil atoms and energy deposition shift deeper into the material, and the damage evolves from a near-surface concentration to a more dispersed deep-layer profile.

(2) Under all energy conditions, at least 53% of the total energy of the incident ions is dissipated through vacancy formation, inducing internal defects within the crystal. Phonon-mediated energy dissipation serves as the primary pathway for converting kinetic energy into internal energy. Additionally, the total number of displacements equals the sum of vacancies and replacement collisions.

(3) Implantation of 300 keV phosphorus ions at a 7° incidence angle produced a uniform amorphous modified layer with a thickness of 461.18 nm. The resulting microstructural changes within the crystal increased the material's absorption at 1070 nm from 0.39% to 4.74%, enabling the application of LADM techniques.

(4) LADM significantly enhances the ductile cutting performance of modified sapphire. With increasing laser power, the extent of the ductile region expands, while the size and density of brittle fracture defects on the groove surface decrease. At 16 W, the ductile–brittle transition depth increased to 450.67 nm, representing a 51.57% improvement over conventional cutting.

Overall, this study provides valuable insights into the interaction mechanisms between ions and the target material, as well as the material removal mechanisms during LADM of implanted sapphire. It also demonstrates the great potential of integrating ion implantation surface modification with ultraprecision machining to enhance the ductile cutting performance of extremely hard and brittle materials.

Author Contributions: Conceptualization, J.K. and J.X.; methodology, J.K. and K.L.; software, H.M. validation, H.M. and K.L.; formal analysis, K.L.; investigation, J.K. and K.L.; resources, X.C. and J.X.; data curation, J.K. and K.L.; writing—original draft preparation, J.K.; writing—review and editing, X.C. and J.X.; visualization, K.L. and J.C.; supervision, X.C. and J.X.; project administration, X.C. and J.X.; funding acquisition, X.C. and J.X. All authors have read and agreed to the published version of the manuscript.

Funding: This research was funded by State Key Laboratory of Intelligent Manufacturing Equipment and Technology Open Fund (grant number IMETKF2023008), Open Research Fund of Hubei Key Laboratory of Modern Manufacturing Quality Engineering (grant number KFJJ-2022006), Doctoral Research Initiation Fund (grant number XJ2024001101) and the National Natural Science Foundation of China (grant number 52225506).

Data Availability Statement: The original contributions presented in the study are included in the article, further inquiries can be directed to the corresponding author.

Conflicts of Interest: The authors declare no conflicts of interest.

References

1. Merberg, G.N.; Harrington, J.A. Optical and Mechanical Properties of Single-Crystal Sapphire Optical Fibers. *Appl. Opt.* **199**3, *32*, 3201–3209. [CrossRef] [PubMed]
2. Yang, D.; Thomas, M.E.; Kaplan, S.G. Measurement of the Infrared Refractive Index of Sapphire as a Function of Temperature. In Proceedings of the Window and Dome Technologies and Materials VII, Orlando, FL, USA, 7 September 2001; SPIE: Bellingham WA, USA, 2001; Volume 4375, pp. 53–63.
3. Voloshin, A.V.; Litvinov, L.A.; Slyunin, E.V. The Influence of Sapphire Crystallographic Orientation on Surface Roughness Achievable by Diamond Abrasive Machining. *J. Superhard Mater.* **2013**, *35*, 56–59. [CrossRef]
4. Locher, J.; Jones, C.; Bates, H.; Rioux, J. Characteristics of Thick (>12 mm) Class225 EFG Sapphire Sheet for IR Window Applications. In Proceedings of the Window and Dome Technologies and Materials XI, Orlando, FL, USA, 11 May 2009; SPIE Bellingham, WA, USA, 2009; Volume 7302, pp. 21–25.
5. Wun, J.-M.; Lin, C.-W.; Chen, W.; Sheu, J.-K.; Lin, C.-L.; Li, Y.-L.; Bowers, J.E.; Shi, J.-W.; Vinogradov, J.; Kruglov, R.; et al. GaN-Based Miniaturized Cyan Light-Emitting Diodes on a Patterned Sapphire Substrate With Improved Fiber Coupling for Very High-Speed Plastic Optical Fiber Communication. *IEEE Photonics J.* **2012**, *4*, 1520–1529. [CrossRef]
6. Savage, J.A. Preparation and Properties of Hard Crystalline Materials for Optical Applications—A Review. *J. Cryst. Growth* **1991**, *113*, 698–715. [CrossRef]
7. Wang, Y.; Liang, Z.; Zhao, W.; Wang, X.; Wang, H. Anisotropic Cutting Mechanisms on the Surface Quality in Ultra-Precision Machining of R-Plane Sapphire. *Appl. Surf. Sci.* **2023**, *622*, 156868. [CrossRef]
8. Ma, Y.; Cao, L.; Hang, W.; Zhang, T.; Yuan, J. Crystallographic Orientation Effect on the Incipient Plasticity and Its Stochastic Behavior of a Sapphire Single Crystal by Spherical Nanoindentation. *Ceram. Int.* **2020**, *46*, 15554–15564. [CrossRef]
9. Bum Kwon, S.; Min, S. Characteristics of Force Generation on C-, R-, A- and M- Planes of Single-Crystal Sapphire during Ultra-Precision Machining. *Manuf. Lett.* **2022**, *33*, 349–354. [CrossRef]
10. Wang, J.; Guo, B.; Zhao, Q.; Zeng, Z.; Zhai, W.; Chen, H.; Sun, J. Investigation into the Anisotropy of Cross-Grinding Surface Quality in C- and M-Planes of Sapphire. *Proc. Inst. Mech. Eng. Part B J. Eng. Manuf.* **2019**, *233*, 44–54. [CrossRef]
11. Maas, P.; Mizumoto, Y.; Kakinuma, Y.; Min, S. Anisotropic Brittle-Ductile Transition of Monocrystalline Sapphire during Orthogonal Cutting and Nanoindentation Experiments. *Nanotechnol. Precis. Eng.* **2018**, *1*, 157–171. [CrossRef]
12. Zhang, S.J.; To, S.; Wang, S.J.; Zhu, Z.W. A Review of Surface Roughness Generation in Ultra-Precision Machining. *Int. J. Mach. Tools Manuf.* **2015**, *91*, 76–95. [CrossRef]
13. Bifano, T.G.; Dow, T.A.; Scattergood, R.O. Ductile-Regime Grinding: A New Technology for Machining Brittle Materials. *J. Eng. Ind.* **1991**, *113*, 184–189. [CrossRef]
14. Wang, X.; Zheng, W.; Bao, X.; Wang, S.; Zhao, Q. Investigation of Anisotropic Characteristics in Ultra-Precision Grinding of Sapphire Crystal Planes and Strategies for Suppression. *J. Manuf. Process.* **2025**, *133*, 55–85. [CrossRef]
15. Wang, X.; Ning, Y.; Zheng, W.; Bao, X.; Zhao, Q.; Wang, S. Anisotropic Mechanism of Material Removal and Ductile-Brittle Transition in Sapphire Scratching Based on Acoustic Emission Signal. *J. Mater. Res. Technol.* **2024**, *31*, 2876–2899. [CrossRef]
16. Maslen, E.N.; Streltsov, V.A.; Streltsova, N.R.; Ishizawa, N.; Satow, Y. Synchrotron X-Ray Study of the Electron Density in α-Al_2O_3 *Acta Cryst. B* **1993**, *49*, 973–980. [CrossRef]
17. Cao, L.; Zhang, X.; Yuan, J.; Guo, L.; Hong, T.; Hang, W.; Ma, Y. Study on the Influence of Sapphire Crystal Orientation on Its Chemical Mechanical Polishing. *Appl. Sci.* **2020**, *10*, 8065. [CrossRef]
18. Kwon, S.B.; Min, S. Studying Mechanism of Anisotropic Crack Generation on C-, R-, A-, and M-Planes of Sapphire during Ultra-Precision Orthogonal Cutting Using a Visualized Slip/Fracture Activation Model. *Nanotechnol. Precis. Eng.* **2024**, *7*, 043006 [CrossRef]
19. Fang, F.Z.; Chen, Y.H.; Zhang, X.D.; Hu, X.T.; Zhang, G.X. Nanometric Cutting of Single Crystal Silicon Surfaces Modified by Ion Implantation. *CIRP Ann.* **2011**, *60*, 527–530. [CrossRef]
20. Wang, J.; Zhang, X.; Fang, F.; Chen, R. Diamond Cutting of Micro-Structure Array on Brittle Material Assisted by Multi-Ion Implantation. *Int. J. Mach. Tools Manuf.* **2019**, *137*, 58–66. [CrossRef]
21. Skuratov, V.A.; Zagorski, D.L.; Efimov, A.E.; Kluev, V.A.; Toporov, Y.P.; Mchedlishvili, B.V. Swift Heavy Ion Irradiation Effect on the Surface of Sapphire Single Crystals. *Radiat. Meas.* **2001**, *34*, 571–576. [CrossRef]
22. Kabir, A.; Meftah, A.; Stoquert, J.P.; Toulemonde, M.; Monnet, I. Amorphization of Sapphire Induced by Swift Heavy Ions: A Two Step Process. *Nucl. Instrum. Methods Phys. Res. Sect. B Beam Interact. Mater. At.* **2008**, *266*, 2976–2980. [CrossRef]
23. Pelaz, L.; Marqués, L.A.; Barbolla, J. Ion-Beam-Induced Amorphization and Recrystallization in Silicon. *J. Appl. Phys.* **2004**, *96*, 5947–5976. [CrossRef]
24. Xu, Z.W.; Fang, F.Z.; Fu, Y.Q.; Zhang, S.J.; Han, T.; Li, J.M. Fabrication of Micro/Nano-Structures Using Focused Ion Beam Implantation and XeF2 Gas-Assisted Etching. *J. Micromech. Microeng.* **2009**, *19*, 054003. [CrossRef]

5. Wang, J.; Chen, R.; Zhang, X.; Fang, F. Study on Machinability of Silicon Irradiated by Swift Ions. *Precis. Eng.* **2018**, *51*, 577–581. [CrossRef]
6. To, S.; Wang, H.; Jelenković, E.V. Enhancement of the Machinability of Silicon by Hydrogen Ion Implantation for Ultra-Precision Micro-Cutting. *Int. J. Mach. Tools Manuf.* **2013**, *74*, 50–55. [CrossRef]
7. He, Y.; Xiao, G.; Liu, Z.; Ni, Y.; Liu, S. Subsurface Damage in Laser-Assisted Machining Titanium Alloys. *Int. J. Mech. Sci.* **2023**, *258*, 108576. [CrossRef]
8. Zhao, B.; Xu, J.; Chen, G.; Zhai, C.; Zhou, F.; Li, J. Study on Chip Formation and Surface Integrity of Nickel-Based Superalloy Inconel 718 by Laser Assisted Micro-Cutting. *J. Manuf. Process.* **2024**, *113*, 34–46. [CrossRef]
9. Du, H.; Yip, W.S.; Sun, W.; Kang, C.; To, S. In-Situ Laser-Assisted Ultraprecision Cutting of WC-Co Cemented Carbide for Creating Microstructure Arrays. *Ceram. Int.* **2023**, *49*, 38698–38707. [CrossRef]
0. Mohammadi, H.; Ravindra, D.; Kode, S.K.; Patten, J.A. Experimental Work on Micro Laser-Assisted Diamond Turning of Silicon (111). *J. Manuf. Process.* **2015**, *19*, 125–128. [CrossRef]
1. Ke, J.; Chen, X.; Liu, C.; Zhang, J.; Yang, H.; Xu, J. Enhancing the Ductile Machinability of Single-Crystal Silicon by Laser-Assisted Diamond Cutting. *Int. J. Adv. Manuf. Technol.* **2022**, *118*, 3265–3282. [CrossRef]
2. Liu, C.; Ke, J.; Yin, T.; Yip, W.S.; Zhang, J.; To, S.; Xu, J. Cutting Mechanism of Reaction-Bonded Silicon Carbide in Laser-Assisted Ultra-Precision Machining. *Int. J. Mach. Tools Manuf.* **2024**, *203*, 104219. [CrossRef]
3. Ke, J.; Chen, X.; Liu, C.; Xu, G.; She, Z.; Zhang, J.; Xu, J. Ultra-Precision Cutting Characteristics of Binderless Tungsten Carbide by in-Heat-Process Laser-Assisted Diamond Machining. *Int. J. Refract. Met. Hard Mater.* **2023**, *115*, 106311. [CrossRef]
4. You, K.; Fang, F.; Yan, G. Surface Generation of Tungsten Carbide in Laser-Assisted Diamond Turning. *Int. J. Mach. Tools Manuf.* **2021**, *168*, 103770. [CrossRef]
5. Zhang, J.; Zheng, Z.; Huang, K.; Lin, C.; Huang, W.; Chen, X.; Xiao, J.; Xu, J. Field-Assisted Machining of Difficult-to-Machine Materials. *Int. J. Extrem. Manuf.* **2024**, *6*, 032002. [CrossRef]
6. Li, Y.; Xing, Y.; Sun, L.; Li, D.; Yin, T.; Wang, J.; Yip, W.S.; To, S. Analytical Thermal Modeling for Conventional and In-Situ Laser Assisted Turning. *Int. J. Mech. Sci.* **2025**, *287*, 109919. [CrossRef]
7. Bauer, P.; Golser, R.; Semrad, D.; Maier-Komor, P.; Aumayr, F.; Arnau, A. Influence of the Chemical State on the Stopping of Protons and He-Ions in Some Oxides. *Nucl. Instrum. Methods Phys. Res. B Beam Interact. Mater. At.* **1998**, *136–138*, 103–108. [CrossRef]
8. Porter, L.E. Analyses of Stopping Power Measurements for 0.90–2.50 MeV Protons and Deuterons Traversing Al2O3 Targets. *Nucl. Instrum. Methods Phys. Res. B Beam Interact. Mater. At.* **2000**, *170*, 35–38. [CrossRef]
9. Shulga, V.I. Note on the Artefacts in SRIM Simulation of Sputtering. *Appl. Surf. Sci.* **2018**, *439*, 456–461. [CrossRef]
0. Hofsäss, H.; Zhang, K.; Mutzke, A. Simulation of Ion Beam Sputtering with SDTrimSP, TRIDYN and SRIM. *Appl. Surf. Sci.* **2014**, *310*, 134–141. [CrossRef]
1. Yoshimura, S.; Hine, K.; Kiuchi, M.; Hashimoto, J.; Terauchi, M.; Honda, Y.; Nishitani, M.; Hamaguchi, S. Experimental Evaluation of CaO, SrO and BaO Sputtering Yields by Ne^+ or Xe^+ Ions. *J. Phys. D Appl. Phys.* **2011**, *44*, 255203. [CrossRef]
2. Ke, J.; Zhang, J.; Chen, X.; Liu, C.; Long, G.; Sun, H.; Xu, J. Investigation on the Material Removal Mechanism in Ion Implantation-Assisted Elliptical Vibration Cutting of Hard and Brittle Material. *Int. J. Mach. Tools Manuf.* **2024**, *203*, 104220. [CrossRef]
3. Egeland, G.W.; Valdez, J.A.; Maloy, S.A.; McClellan, K.J.; Sickafus, K.E.; Bond, G.M. Heavy-Ion Irradiation Defect Accumulation in ZrN Characterized by TEM, GIXRD, Nanoindentation, and Helium Desorption. *J. Nucl. Mater.* **2013**, *435*, 77–87. [CrossRef]
4. Yoon, H.-S.; Kwon, S.B.; Nagaraj, A.; Lee, S.; Min, S. Study of Stress Intensity Factor on the Anisotropic Machining Behavior of Single Crystal Sapphire. *CIRP Ann.* **2018**, *67*, 125–128. [CrossRef]
5. Yoon, H.-S.; Kwon, S.B.; Kim, J.-H.; Ahn, S.-H.; Min, S. Effects of Surface Coating Materials on Cutting Forces and Ductile-to-Brittle Transition in Orthogonal Cutting of Monocrystalline Sapphire. *J. Manuf. Process.* **2022**, *84*, 375–382. [CrossRef]
6. Yoon, H.-S.; Kwon, S.B.; Nagaraj, A.; Min, S. Effect of the Initial-Flaw on Crack-Propagation in Two-Step Cutting of Monocrystalline Sapphire. *J. Manuf. Process.* **2020**, *56*, 1211–1218. [CrossRef]

micromachines

Article

An Effective Hybrid Rescheduling Method for Wafer Chip Precision Packaging Workshops in Complex Manufacturing Environments

Ziyue Wang [1,2], Weikang Fang [1,*] and Yichen Yang [3]

1 School of Smart Manufacturing, Jianghan University, Wuhan 430056, China; wangziyue@jhun.edu.cn
2 School of Mechanical Science and Engineering, Huazhong University of Science and Technology, Wuhan 430074, China
3 Marine Design and Research Institute of China, Shanghai 200011, China; yicen1994@163.com
* Correspondence: fangweikangjh@163.com

Abstract: With the continuous development of semiconductor manufacturing technology and information technology, the sizes of wafer chips are becoming smaller and the variety is increasing, which has put forward high requirements for wafer chip precision manufacturing and packaging workshops. On the one hand, the market demand for multiple varieties and small batches will increase the difficulty of scheduling. On the other hand, the complex manufacturing environment brings various types of dynamic events that will disrupt production plans. Accordingly, this work researches the wafer chip precision packaging workshop rescheduling problem under events of machine breakdown, emergency order inserting and original order modification. Firstly, the mathematical model for the addressed problem is established, and the rolling horizon technology is adopted to deal with multiple dynamic events. Then, a hybrid algorithm combining an improved firefly optimization framework and variable neighborhood search strategy is proposed. The population evolution mechanism is designed based on the location-updating law of fireflies in nature. The variable neighborhood search is applied for avoiding local optima and sufficiently exploring in the neighborhood. At last, the test results of comparative experiments and engineering cases indicate that the proposed method is effective and stable and is superior to the current advanced algorithms.

Keywords: wafer chip; precision manufacturing and packaging; rescheduling; lot streaming; dynamic events

1. Introduction

Semiconductor chips, with their advantages of powerful functions and small volume, are widely used in fields such as the manufacturing industry, household appliances, medical devices, vehicles, and electrical products [1]. At present, semiconductor chips are constantly developing towards the directions of large batch and diversification [2]. Hence, higher requirements have been put forward for chip precision manufacturing, especially for the chip precision packaging production lines that are located in the middle and lower reaches of production chain [3]. Furthermore, the manufacturing system of chip precision packaging workshops is required to aim at high production efficiency, low operating costs, and timely delivery to adapt to complex and ever-changing market demands. Therefore, it poses significant challenges to the semiconductor chip precision packaging production line scheduling problem (SCPPLSP), but there are currently relatively few works on this issue.

 https://doi.org/10.3390/mi1612140

In the semiconductor chip packaging process, wafers are processed and transported in batch forms. In order to reduce transportation costs and equipment adjustment time caused by frequent switching for different types of chips, chips of same type tend to be grouped in several batches for centralized processing, instead of being processed one by one, which is known as the lot streaming [4]. Unfortunately, under the production mode of large batches and multiple varieties, a large number of batches will drastically increase the scale of the solution space for scheduling problems [5]. Some scholars have concluded through theoretical deduction and extensive experiments that reasonable batching strategy can effectively prune the solution space for the lot streaming scheduling problem [6]. However, these existing batching strategies were specifically designed for the problems with specific characteristics and scales, which are not the primary way to solve engineering problems. An intelligent optimization algorithm is currently one of the effective ways to solve complex scheduling problems but it has low search efficiency and poor solution quality for the lot streaming scheduling problem. Therefore, it is urgent to develop reasonable and efficient intelligent optimization algorithms based on the problem characteristics to improve the quality and efficiency for the addressed SCPPLSP.

The semiconductor chip packaging process is often affected by multiple dynamic uncertainties [7]. For instance, there are many types and quantities of machines or processing units required for the chip packaging process, and hence machinery breakdown events often occur. As another example, affected by the shortage of chips in the semiconductor and electronic manufacturing market and the diversification of demands, events such as emergency orders and order cancelations often occur. The above dynamic uncertainty interferences existing in the chip packaging production line are mostly sudden events, and the rescheduling mode is commonly applied for this type of dynamic scheduling problems [8]. However, the existing studies on rescheduling problems mainly focus on the single dynamic event. Therefore, it is urgently needed to analyze the impact of dynamic event and propose rescheduling methods for the scheduling problems under multiple perturbations.

Based on the above analysis, the SCPPLSP under the multiple dynamic events will be addressed in this paper. For the scheduling optimization problems, establishing the mixed-integer linear programming (MILP) model is the foundation for analyzing the problems' characteristics and solving combinatorial optimization problems [9]. Hence, in this work, the MILP model for the SCPPLSP is established first, and then the rescheduling method based on rolling horizon technology [10] is given to deal with the multiple different dynamic events. Firefly algorithm (FA) is designed based on simulating the flashing behaviors and movement law of fireflies in nature, and has the advantages of simple encoding, strong global search ability, flexible framework for easy improvement, and easy tuning for parameters [11]. Compared to other swarm intelligence optimization algorithms, the improved FA frameworks are more commonly used to solve complex scheduling optimization problems in practical engineering [12–14]. Consequently, enlightened by the design concept of standard FA, this research constructed a new improved FA framework for the SCPPLSP. To avoid local optima and strengthen global searching, the variable neighborhood descent (VND) strategy is added to the proposed improved FA framework.

The rest of this work is arranged as below. Section 2 summarizes the literatures in aforesaid fields. Section 3 characterizes the SCPPLSP and establishes the MILP model for it. In Section 4, the presented hybrid FA based on VND (HFA-VND) is explicated in detail. In Section 5, the comparison experiments and the results analysis are presented, and an engineering case verification is given. Section 6 summarizes the entire text.

2. Related Works

2.1. Semiconductor Manufacturing System

The semiconductor manufacturing system scheduling problem is an academic hotspot, and the scholars in the related fields have obtained certain study achievements [15,16]. Yang et al. [17] focused on the issues of wafer reentrance and residency time constraints. Then, they carried out the schedulability analysis on the different strategies, derived schedulability conditions, and proposed a scheduling algorithm to obtain the optimal scheduling scheme. Wang et al. [18] considered the correlation between wafer lots and production cycles in reentrant processes and put forward a fuzzy hierarchical model to adapt the production cycle of each beach. There are a variety of different performance indicators for semiconductor manufacturing systems, such as work-in-progress quantity, wafer residency time, equipment utilization rate, production cycle, operating cost, total movement volume, on-time delivery rate, energy consumption, and so on [19]. To improve the equipment utilization rate and throughput of semiconductor manufacturing systems, Siebert et al. [20] provided a batch scheduling strategy for uncertain environments. Calmels [21] addressed the multi-objective conflict problem in semiconductor manufacturing and presented a mixed-integer programming formulation along with a corresponding heuristic local search strategy. With the improvement of information technology, the analytical methods based on industrial big data technology have gradually been applied for modeling and optimization of the semiconductor manufacturing system [22–25]. Park et al. [26] addressed the large-scale production issue in semiconductor manufacturing by adopting a deep reinforcement learning approach and introducing a new state representation to adapt to changes in available resources and production requirements. Lee et al. [27] focused on the reentrant phenomenon in semiconductor manufacturing, and improved the state, action, and reward mechanisms to optimize the scheduling strategy. Although, the above studies have covered issues of reentrant processing, multi-objective optimization, uncertain disturbances, and so on, there is almost no research on SCPPLSP under the interference of multiple dynamic events.

2.2. Lot Streaming Scheduling

The scheduling problem with lot streaming is an integrated optimization problem of batching and scheduling, which is the NP-hard problem [28–31]. Currently, based on different batching methods, the related studies can be divided into three modes: equal-size uniform batching [32], unequal-size uniform batching [33], and variable sub-lot [34]. All of these modes have achieved certain research results. Zhang et al. [32] proposed an improved migratory bird optimization algorithm (MBO) to solve the lot streaming hybrid flowshop scheduling problem (HFSP). Wang et al. [35] focused on the integrated problem of batch processing and lot streaming in a two-stage HFSP and adopted a heuristic algorithm to minimize the total weighted completion time. Nejati et al. [36] studied the HFSP with lot streaming under multi-stage unequal-size lots with shift constraints. Considering the scenario of interleavable sublots, Nejati et al. [37] studied a two-stage assembly line with lot streaming and unequal-size lots. They took the makespan as the optimization objective, and applied the genetic algorithm (GA) and simulated annealing (SA) algorithm, respectively. For a special form of hybrid flow shop scheduling problem with lot streaming and multi-stage unequal-size lots, Lalitha et al. [38] established the MILP model and proposed a related heuristic rule for solving. For the lot streaming scheduling problem, Novas [39] designed a novel constraint method and determined the number of sublots for all lots in the scheduling scheme. A large number of batch divisions will greatly enlarge the size of solution space [6,40], and reasonable batching strategy can effectively improve the solving efficiency. Vivek et al. [41] used a heuristic method to evaluate 11 batching strategies and obtain the optimal batching method. However, the batching strategy is usually specifically designed for the problems with specific characteristics and scales. Hence, proposing an efficient optimization algorithm is an effective way to solving scheduling optimization problem with lot streaming.

2.3. Dynamic Scheduling

The complex production environment and diverse market demands [42] will disrupt the scheduling plans formulated by the manufacturing system. The scholars dedicated to dynamic scheduling problems mainly focus on scheduling methods, such as rescheduling [43], inverse scheduling [44], and robust scheduling [45–47]. Rescheduling mode is often adopted for the unexpected dynamic events that have significant impact on production plans. Zhang et al. [10] presented a rescheduling method based on a multi-objective MBO for HFSP. He et al. [48] provided a rescheduling model for hybrid flow shop to address the problem of urgent order insertion and used the NSGA-III algorithm to solve it. Inverse scheduling mode involves adjusting controllable resources and processing parameters to meet production requirements instead of changing the production plan, which is suitable for the unexpected dynamic events that have small-scale impact on production plans [49]. Mou et al. [50] adopted an inverse scheduling mode and studied a hybrid algorithm for the energy-efficient scheduling problem. Zhang et al. [51] proposed an efficient hybrid integer and a categorical particle swarm optimization algorithm for the multi-mode multi-project inverse scheduling problem in the turbine assembly workshop. For random uncertain disturbances, robust scheduling strategies are often employed. Qiao et al. [52] proposed a robust real-time scheduling strategy that combined a real-time controller with offline scheduling. To tackle the problem of uncertain processing times in semiconductor manufacturing systems, Liu et al. [53] proposed a three-stage multi-objective robust optimization method to enhance system stability. Preventive maintenance strategy is one of the robust scheduling methods, which is used to handle dynamically predictable events. Lee et al. [54] designed a preventive maintenance optimization model to prevent the impact of downtime events on semiconductor manufacturing systems. Obviously, there have been certain studies on the dynamic scheduling problems under a single dynamic event, but the related research on multiple types of uncertain events is paid little attention, especially for the scheduling problem in semiconductor chip packaging production lines.

3. Problem Formulation

3.1. Problem Description

SCPPLSP can be equivalent to an HFSP with the lot streaming processing mode and the interference of multiple dynamic events. In this research, the most common dynamic events in chip packaging production lines, namely machine breakdown, emergency order inserting, and original order modification, are considered in SCPPLSP. Then, the SCPPLSP under multiple dynamic events can be described as follows: the chips of several types in the order need to be continuously processed in several processing stages, and the processing steps for all the chips are the same. In each processing stage, there are several same processing units or machines available for selection. The chips of each type are divided into sub-lots with the same number. Each sub-lot that has been completed at a certain processing stage can be immediately carried to the next process stage, while the different sub-lots can be processed simultaneously at different stages, which accelerates the operation of the production line and improves production efficiency. In SCPPLSP, the sub-lots are treated as the smallest processing units for scheduling. SCPPLSP aims to minimize the makespan as the optimization objective. Here, the following hypothesis settings are given:

- To avoid the frequent switching, all sub-lots of each type of chip can only be processed continuously by the same processing unit or machine at each stage;
- All the processing unit can only deal with one sub-lot simultaneously, and each sub-lot can only be processed by one processing unit;
- There is no waiting time limit for each sub-lot between the adjacent processing stages;
- The capacity of the buffer between the adjacent processing stages is usually set to be much larger than the number of chips in process, so its impact on production plan and rhythm can be approximately ignored;
- The setup times can be approximated to be sequence independent and have been calculated within the processing times;
- The transportation time between adjacent processing stages has been calculated in the processing time;
- Machine breakdown events randomly occur at any time, and the maintenance time can be accurately predicted;
- Emergency order inserting events can randomly occur randomly at any time, and a certain number of sub-lots will be added to the production plan;
- Original order modifications can randomly occur randomly at any time, and a certain number of sub-lots will be removed from the production plan.

3.2. Rescheduling Mode

The dynamic events described in this work, namely machine breakdown, emergency order inserting, and original order modification, can be treated as sudden events, and it is generally difficult to accurately predict the exact time when these dynamic events will occur. Therefore, the rescheduling mode is adopted here. The rescheduling mode can be described as follows: when a dynamic event occurs, the unprocessed operations are rescheduled for the optimization of performance indicators. The rolling horizon method is commonly used for the rescheduling problem, which can divide the timeline into several parts [13]. In this research, different types of dynamic events occur at different times, so at each time when a dynamic event occurs, namely the rescheduling time, all the operations of each sub-lot need to be analyzed and judged based on the rolling horizon theory.

After a dynamic event occurs, all operations of each sub-lot will be divided into different state sets based on their processing status. State set 1 includes the processed operations and the ones that have been started on the non-faulty processing units. State set 2 includes the unprocessed operations. The operation being processed on the faulty processing unit or machines belongs to State set 3, which can be processed continuously after the faulty processing unit is restored. For the urgently inserted order, all the operations of the chips in the order should be added to the State set 2, which will be treated as a new chip type in rescheduling plan. For the modified original order, all the operations of the canceled chips will be canceled from the rescheduling plan. Hence, only the operations in State set 2 should be rearranged at the rescheduling time.

Here, a simple instance is given to show the rescheduling mode under three dynamic events described above, as shown in Figure 1. (j,e) represents the sub-lot e of the chip type j. At rescheduling time 1, machine 2 at the first processing stage breaks down. The operations (1,1), (1,2), (1,3) on the faulty machine, which are in the yellow box, belong to State set 3, while the operations (2,1), (4,1), (4,2) at stage 1 and the operations (2,1), (2,2), (4,1), (4,2) at stage 2 belong to State set 1. The other operations belong to State set 2. At rescheduling time 2, the emergency order is inserted and one sub-lot of chip type 3 is added, namely operation $(3',1)$, which is in the green box and is treated as a new type for rescheduling. At rescheduling time 3, the original order is modified and the operation (5,2) is removed, as shown in the blue box. At this time, the operation $(3',1)$ at stage 1, the operations $(3',1)$,

(5,1) at stage 2, and the operations of (3,1), (3′,1), (5,1) at stage 3 belong to State set 2, which should be rescheduled.

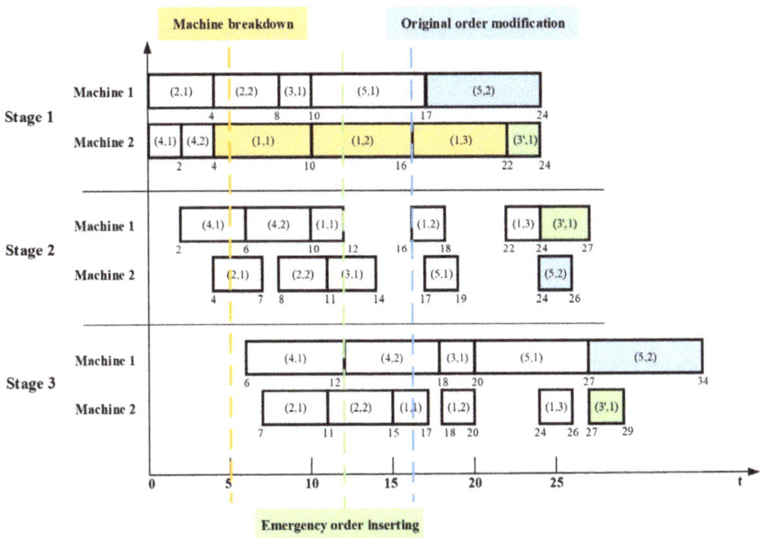

Figure 1. The rescheduling mode.

3.3. Mathematical Model

According to the problem description of SCPPLSP and the rescheduling mode based on the rolling horizon method, the MILP model for SCPPLSP is established. Firstly, the related symbols are defined as in Abbreviations.

Then, give the MILP model for SCPPLSP:

- **Objective:**

$$f = tC_{max} = \max(tC_1, tC_2, tC_3, \ldots, tC_n) \tag{1}$$

- **Subject to:**

$$\sum_{j=1}^{m_k} bD'_{i,j,k} = 1 \; \forall i \in \boldsymbol{J},\; k \in \boldsymbol{M},\; j \in \boldsymbol{M_k} \tag{2}$$

$$tC'_{i,e,k} - tS'_{i,e,k} = tP_{i,k} \quad \forall O_{i,e,k} \in \textbf{\textit{State}}\ 2 \tag{3}$$

$$tS'_{i,e,k+1} - tC'_{i,e,k} \geq 0 \; \forall O_{i,e,k} \in \textbf{\textit{State}}\ 2 \tag{4}$$

$$tS'_{i,e+1,k} - tC'_{i,e,k} \geq 0 \; \forall O_{i,e,k} \in \textbf{\textit{State}}\ 2 \tag{5}$$

$$bS'_{i\prime,i,k} + bS'_{i,i\prime,k} \leq 1 \; \forall O_{i,e,k}, O_{i\prime,e\prime,k} \in \textbf{\textit{State}}\ 2 \tag{6}$$

$$bS'_{i,i\prime,k} + bS'_{i\prime,i,k} \leq bD'_{i,j,k} + bD'_{i\prime,j,k} \; \forall O_{i,e,k}, O_{i\prime,e\prime,k} \in \textbf{\textit{State}}\ 2 \tag{7}$$

$$bD'_{i,j,k} + bD'_{i\prime,j,k} - 1 \leq bS'_{i,i\prime,k} + bS'_{i\prime,i,k} \; \forall O_{i,e,k}, O_{i\prime,e\prime,k} \in \textbf{\textit{State}}\ 2 \tag{8}$$

$$tS'_{i,1,k} - tC'_{i\prime,l_{i\prime},k} + Q \times (3 - bS'_{i,i\prime,k} - bD'_{i,j,k} - bD'_{i\prime,j,k}) \geq 0\ \forall O_{i,e,k}, O_{i\prime,e\prime,k} \in \textbf{\textit{State}}\ 2 \quad (9)$$

$$tS'_{i,e,k} = tS_{i,e,k}\ tC'_{i,e,k} = tC_{i,e,k}\ \forall O_{i,e,k} \in \textbf{\textit{State}}\ 1 \quad (10)$$

$$bD'_{i,j,k} = bD_{i,j,k}\ bS'_{i,i\prime,k} = bS_{i,i\prime,k}\ \forall O_{i,e,k}, O_{i\prime,e\prime,k} \in \textbf{\textit{State}}\ 1 \cup \textbf{\textit{State}}\ 3 \quad (11)$$

$$tS'_{i,e,k} = tS_{i,e,k}\ tC'_{i,e,k} = tC_{i,e,k} + tB - tR\ \forall O_{i,e,k} \in \textbf{\textit{State}}\ 3 \quad (12)$$

$$bD'_{i,j,k} \in \{0,1\}\ bS'_{i,i\prime,k} \in \{0,1\}\ \forall O_{i,e,k}, O_{i\prime,e\prime,k} \in \textbf{\textit{State}}\ 2 \quad (13)$$

Equation (1) states that the optimization objective of the MILP model for SCPPLSP is the makespan. Constraints (2)–(9) provide the constrained condition for SCPPLSP in rescheduling. Constraints (10) and (11) state that the rescheduling plans of the operations in State set 1 are the same as the original plans of them. Constraint (12) states that, in the rescheduling plan, the operation belonging to State set 3 will be processed on the faulty machine continuously after the faulty machine is restored. Constraint (13) provides the value ranges for the two decision variables in rescheduling, respectively.

4. Hybrid Firefly Algorithm Based on Variable Neighborhood Descent

4.1. Algorithm Framework

As described in Section 1, FA has the advantages of simple encoding, strong global search ability, flexible framework for easy improvement, and easy tuning for parameters, which is more commonly improved to solve actual complex scheduling problems in production shop. Accordingly, an improved FA framework is proposed here to solve SCPPLSP. The design concept of standard FA originates from the movement patterns of firefly in nature, where the darker individuals will move closer to the brighter ones. Inspired by the FA concept, in our improved framework, the suboptimal individuals move toward more optimal ones by designing movement operator, while the random perturbation is introduced to enhance the population diversity. To avoid local optima and strengthen global searching, the local search operation based on VND is adopted in above algorithm framework. The algorithm flow of HFA-VND is shown in Algorithm 1:

Algorithm 1 The algorithm flow of HFA-VND

1. Set the population size *ps* and the maximum number of iterations for VND *maxtsIter;*
2. **Population initialization**: Generate initial population containing *ps* individuals;
3. **Individuals movement**: In stage β, compare all the individuals pairwise and make suboptimal individuals move toward more optimal ones, while in stage α, the random perturbations are used to avoid premature entry into local optima;
4. **Local search**: Select several optimal individuals to conduct the local search operation to ensure sufficient mining in their neighborhood;
5. Go to step 6 if the termination condition is met; go to step 3 otherwise;
6. Output the results.

4.2. Individual Description

As described in Section 3.1, in the SCPPLSP, all sub-lots can only be processed by one processing unit, and all the chips have the same processing sequence. Therefore, we adopt an encoding mechanism based on the sequence of chip types, which represents the processing order of chip types in the initial stage and has been effectively used for solving HFSP. The encoding sequence can be represented as: $\{E(1), E(2), \ldots, E(n)\}$, where $E(i)$ is the index of the chip type.

According to the problem characteristics of SCPPLSP and the above encoding structure, it is required that the decoding mechanism should solve the following two sub-problems: one is the sequence of chip types at other processing stages, and the other is the machine selection for different chip types at each stage. In order to shorten the waiting times of chips and the idle times of machines, two heuristic rules are adopted as below:

- The "first come, first processed for sub-lot" rule: a certain sub-lot of a certain chip type that has been processed at a certain stage does not need to wait for the other subsequent sub-lots of this chip type, and can directly start to be processed at the subsequent stage;
- The "first idle machine, first processing" rule: the sub-lot that needs to be processed should be assigned on the first idle machine at this stage at this time.

Based on the above analysis, the decoding mechanism can be described in Algorithm 2:

Algorithm 2 The algorithm flow of the decoding mechanism

1. All the operations of all the sub-lots should be divided into three state sets based on the rolling horizon technology;
2. Find the earliest processing stage s′ ($1 \leq s' \leq m$) among all the operations in State set 2;
3. Record the number n_k ($k \in \{s', s' + 1 \ldots, m\}$) of the chip types belonging to State set 2 at each stage;
4. Sort the chip types belonging to State set 2 at the stage s' based on the encoding sequence;
5. **For** i **in** $\{1,2,\ldots, n_{s'}\}$ **do**
6. Select the first idle machine, and process all the sub-lots of the chip type i sequentially on this machine;
7. Calculate the processing times of these sub-lots, and use the completion time of the last sub-lot to update the machine idle time;
8. **End**
9. **For** k **in** $\{s' + 1 \ldots, m\}$ **do**
10. Sort the chip types belonging to State set 2 at the stage k according to the ascending order of completion time of the first sub-lot of each type;
11. **For** i **in** $\{1,2,\ldots, n_k\}$ **do**
12. Select the first idle machine, and process all the sub-lots of chip type i sequentially on this machine;
13. Calculate the processing times of these sub-lots, and use the completion time of the last sub-lot to update the machine idle time;
14. **End**
15. **End**

4.3. Population Initialization

To ensure the diversity and superiority of initial individuals, a hybrid initialization strategy based on the integrating random rule and NEH rule is proposed here to generate the initial population. To avoid the long waiting time, two heuristic rules are adopted, respectively, in the NEH rule, namely the longest processing time rule (LPT) and the longest processing time at the first stage rule (LPTF). The hybrid initialization strategy can be described in Algorithm 3:

Algorithm 3 The algorithm flow of the hybrid initialization strategy

1. Randomly generate *ps*-2 individuals;
2. Generate an individual according to the NEH rule based on LPT:
    - Calculate the total processing times of all chip types, and sort the chip types based on their descending sequence;
    - Successively take out the chip types from the above sequence and insert them into all available positions in order to obtain the optimal subsequence;
    - Proceed until the last chip type is inserted into a suitable position, and the optimal individuals are obtained;
3. Generate an individual according to the NEH rule based on LPTF:
    - Calculate the processing times of all chip types at the first stage, and sort the chip types based on their descending sequence;
    - Successively take out the chip types from the above sequence and insert them into all available positions in order to obtain the optimal subsequence;
    - Proceed until the last chip type is inserted into a suitable position, and the optimal individuals are obtained;
4. The above ps individuals are used to form the initial population.

4.4. Individual Movement

According to the design thought of standard FA, the individual movement process includes two stages: in the stage β, the suboptimal individuals move closer to more optimal ones, and in the stage α, the individual is randomly modified to avoid local optima. As described in Section 4.1, the improved FA framework proposed in the paper is designed inspired by the core idea of the standard FA. As an improvement, we do not use the firefly position update equation of standard FA to achieve individual movement. Instead, according to the characteristics of the addressed SCPPLSP and the encoding structure described in Section 4.2, in the stage β of the improved algorithm framework, a position-based crossover (PBX) operator is used to realize the individual movement. As shown in Figure 2, by using the PBX operator for two individuals, the new individual will inherit the excellent genetic fragments from the better individual, which achieves the movement of the suboptimal individual towards the better one. Then, in stage α of the improved algorithm framework, the random perturbation based on swap operator and insertion operator is applied for avoiding local optima, as shown in Figures 3 and 4. In this research, only one operator is selected for random variation for each individual, and here the probability of both above operators being selected is 50%.

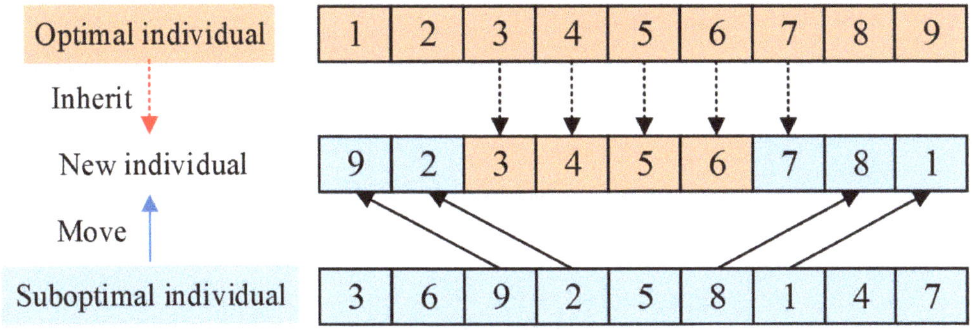

Figure 2. The individual movement strategy based on PBX.

Figure 3. The random perturbation operators: swap operator.

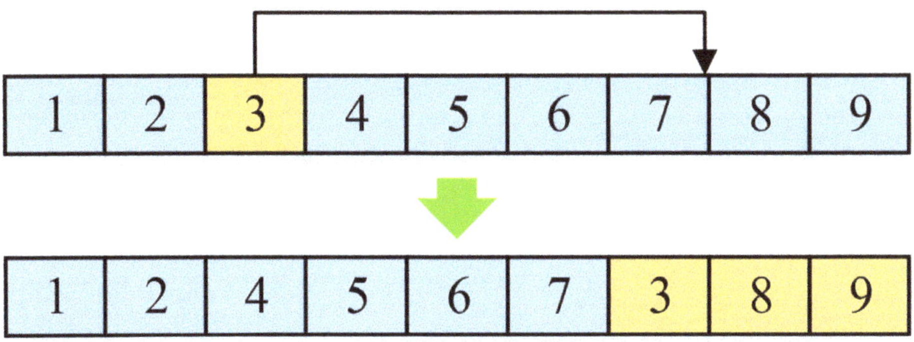

Figure 4. The random perturbation operators: insertion operator.

4.5. Local Search

To avoid local optima and sufficiently explore the neighborhood of the optimal individuals, the VND strategy is adopted and introduced to the improved FA framework. The design concept of VND can be described as follows: if the more optimal individual cannot be mined from the current neighborhood after a certain number of iterations, the next neighborhood will be selected for local search, while if the more optimal individual is mined, the search operation will turn back to the first neighborhood. The neighborhood structures selected for VND in this work are the swap operator and insertion operator described in Section 4.4. The pseudocode of the VND is shown in Algorithm 4:

Algorithm 4 The algorithm flow of VND

```
r = 1, iter = 0;
While (isStop == False)
    X' = N_r(X);
    If f(X') < f(X)
        X'_i = X_i, r = 1, iter = 0;
    Else
        iter++;
    End
    If iter > maxtsIter
        If r < 2
            r++, iter = 0;
        Else
            r = 1, iter = 0, isStop = true;
        End
    End
End
```

5. Experimental Study

5.1. Experimental Design

There has been no benchmark for the addressed SCPPLSP at present, so we selected the test instances randomly generated to evaluate the algorithm performance, which is widely applied for the complex scheduling problem. To simulate the actual processing environment in the semiconductor chip packaging production line, 5 different scales of numbers of the chip types and 2 different scales of numbers of the processing stages are selected to form the test instances of 10 scales. In total, 10 specific test instances are randomly generated for each scale, so a total of 100 test instances are obtained. The occurrence times of these three dynamic events are also randomly valued in the integer interval [1, 99], and the duration of machine breakdown is randomly chosen in the integer interval [20, 99]. The other parameters of these instances are listed in Table 1.

Table 1. The instances parameters.

Parameters	Values
n	{20, 40, 60, 80, 100}
s	{5, 10}
m_k	integer in [5, 10]
$tP_{i,k}$	integer in [10, 20]
l_i	integer in [5, 10]

All the algorithms for comparison tests are realized by Visual C++ language in Visual Studio 2015 on a computer with AMD Ryzen 7 5800H CPU at 3.20 GHz and 16G RAM. The test study in this section will use the best C_{max} and the relative percentage increase (*RPI*) obtained from different algorithms as the references to verify the superiority of HFA-VND. The *RPI* for the algorithm $\boldsymbol{X}$ is formulated as:

$$RPI(\boldsymbol{X}) = (tC_{X-best} - tC_{best})/tC_{best} \times 100\% \quad (14)$$

where $C_{X\text{-}best}$ and C_{best} are the best solutions, respectively, obtained from the algorithm $\boldsymbol{X}$ and all the algorithms. In order to respond promptly to dynamic events with minimal impact on production efficiency, the algorithm execution time limit is set as the algorithm termination condition. Here, the maximum execution time is designed as $10 \times m \times n$ (ms), which can ensure that the production plans will be adjusted within tens of seconds after the dynamic events occur and meet the production demands for SCPPLSP in actual working conditions.

5.2. Algorithm Comparison

To test the performance of HFA-VND, this section introduces the existing advanced algorithms for comparative testing. The introduced algorithms are GA [37], genetic algorithm and tabu search (GATS) [55], genetic algorithm and variable neighborhood search (GAVNS) [56] and MBO [32], which are widely applied for the complex scheduling problem and achieve excellent results.

The population size *ps* and the maximum number of iterations for VND *maxtsIter* are selected through Taguchi experimental testing, which is specially used for the parameter selection and calibration in the field of scheduling optimization. The test results are shown in Figure 5. Seen from Figure 5, population size *ps* and the maximum number of iterations *maxtsIter* are set as 6 and 9, respectively. In GA, GATS and GAVNS, the algorithm termination times are set as the same as HFA-VND. The other parameters of them are also chosen after Taguchi experimental testing. The parameters in MBO are set according to the references [32].

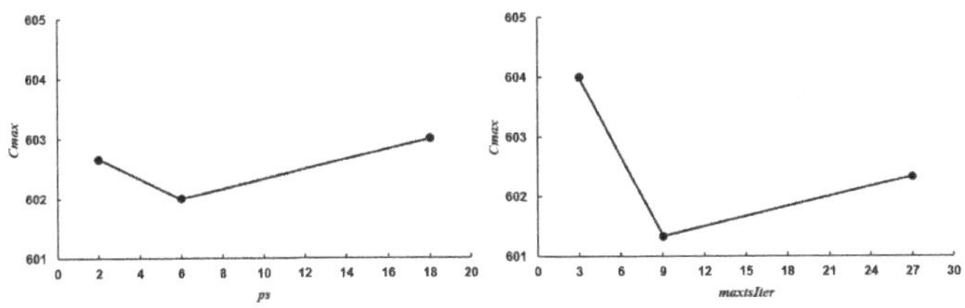

Figure 5. Taguchi test results for HFA-VND.

The 100 test instances generated according to Section 5.1 are used for comparative testing. Under each instance, every algorithm need to be tested 10 times. Then, the best C_{max} and the averages of *RPI* obtained from each algorithm under each test instance are recorded in Table 2. The statistical information (the numbers of best C_{max} or *RPI*, the averages of *RPI* and the variance of *RPI*) of test results for each algorithm is recorded at the bottom of Table 2. The boxplot for the averages of *RPI* obtained from different algorithms under each instance is shown in Figure 6.

Table 2. The comparison experimental results.

Instance	HFA-VND		MBO		GATS		GAVNS		GA	
$m \times n$	C_{max}	*RPI* (%)	C_{max}	*RPI* (%)	C_{max}	*RPI* (%)	C_{max}	*RPI* (%)	C_{max}	*RPI* (%)
20 × 5	**480**	**0.1875**	**480**	0.8125	**480**	0.4583	484	1.4583	486	1.9792
	517	**0.0000**	**517**	0.3482	**517**	0.0193	518	1.1605	521	2.0309
	477	**0.0000**	**477**	0.0629	**477**	**0.0000**	**477**	0.0839	**477**	0.3145
	321	**0.0000**	**321**	0.7788	**321**	**0.0000**	322	2.2741	329	4.1745
	476	**0.0000**	**476**	0.2311	**476**	**0.0000**	**476**	0.5252	481	2.0588
	710	**0.0000**	**710**	0.0845	**710**	**0.0000**	**710**	0.0423	**710**	0.0423
	854	**0.0000**	**854**	**0.0000**	**854**	**0.0000**	**854**	0.0234	**854**	0.0585
	543	**0.0000**	**543**	**0.0000**	**543**	**0.0000**	**543**	0.0368	**543**	0.1105
	372	**0.0000**	**372**	0.6989	**372**	**0.0000**	**372**	0.4839	375	1.3172
	936	**0.0000**	**936**	**0.0000**	**936**	**0.0000**	**936**	0.0107	**936**	0.0214

Table 2. *Cont.*

Instance $m \times n$	HFA-VND		MBO		GATS		GAVNS		GA	
	C_{max}	*RPI* (%)	C_{max}	*RPI* (%)	C_{max}	*RPI* (%)	C_{max}	*RPI* (%)	C_{max}	*RPI* (%)
20 × 10	**1029**	**0.0000**	**1029**	**0.0000**	**1029**	**0.0000**	**1029**	**0.0000**	**1029**	**0.0000**
	848	**0.0000**	**848**	0.1533	**848**	**0.0000**	**848**	0.0825	**848**	0.0708
	772	**0.0000**	**772**	0.1943	**772**	**0.0000**	**772**	0.2073	**772**	0.5829
	1104	**0.0000**	**1104**	**0.0000**	**1104**	**0.0000**	**1104**	**0.0000**	**1104**	0.5072
	966	**0.4037**	973	0.9627	968	0.5901	972	1.9048	974	3.4990
	519	**0.0000**	**519**	**0.0000**	**519**	**0.0000**	**519**	0.3083	**519**	1.0405
	1101	**0.0000**	**1101**	0.1817	**1101**	0.0091	**1101**	0.8629	**1101**	2.6431
	927	**0.0000**	**927**	0.0647	**927**	**0.0000**	**927**	0.1834	**927**	0.7443
	984	**0.0000**	**984**	0.0610	**984**	**0.0000**	**984**	0.0813	**984**	0.2642
	662	**0.0000**	**662**	0.6495	**662**	0.1511	668	2.5680	693	5.7100
40 × 5	**503**	**1.1730**	506	1.7495	508	1.7296	517	5.6461	539	8.8469
	930	**0.1184**	930	0.2476	**929**	0.1830	940	2.9064	963	5.6189
	1379	**0.0000**	**1379**	**0.0000**	**1379**	**0.0000**	**1379**	**0.0000**	**1379**	0.1015
	1667	**0.0000**	**1667**	**0.0000**	**1667**	**0.0000**	**1667**	**0.0000**	**1667**	**0.0000**
	944	**0.0106**	**944**	0.9216	**944**	0.2331	948	1.2606	954	2.3093
	609	**0.0000**	**609**	**0.0000**	**609**	**0.0000**	**609**	0.0821	**609**	0.3777
	601	**0.0000**	602	0.4825	**601**	0.2163	604	0.8319	608	1.5973
	771	**0.1427**	**771**	0.4410	772	0.3891	776	1.8029	786	3.7354
	694	**0.1009**	695	0.6484	**694**	0.1873	702	2.1470	706	3.1844
	800	**0.0000**	**800**	**0.0000**	**800**	**0.0000**	**800**	0.2250	801	0.9250
40 × 10	**1401**	**0.0000**	**1401**	**0.0000**	**1401**	**0.0000**	**1401**	0.5639	**1401**	1.7987
	1541	**0.0000**	**1541**	0.1363	**1541**	0.0454	**1541**	0.2141	1544	0.3504
	1473	**0.0000**	**1473**	0.0815	**1473**	0.0950	**1473**	0.3530	1478	0.5635
	1788	**0.0000**	**1788**	0.0224	**1788**	**0.0000**	**1788**	0.5817	1794	2.0246
	1442	**0.0000**	**1442**	0.0624	**1442**	**0.0000**	**1442**	0.1526	1447	0.9015
	1645	**0.0243**	**1645**	0.2736	**1645**	0.1216	1647	0.3404	1650	0.5410
	1180	**0.1271**	1181	0.1695	1182	0.1695	1190	2.3220	1211	4.5763
	1439	**0.0000**	**1439**	**0.0000**	**1439**	**0.0000**	**1439**	0.2224	**1439**	0.7088
	1666	**0.1200**	**1666**	0.4742	**1666**	0.2221	1681	2.3229	1732	5.0360
	1365	**0.0000**	**1365**	0.1465	**1365**	0.0293	**1365**	0.1612	**1365**	0.4762
60 × 5	**2449**	**0.0000**	**2449**	0.0817	**2449**	**0.0000**	**2449**	**0.0000**	**2449**	0.0204
	987	**0.0101**	988	0.2128	**987**	0.1621	**987**	0.6586	996	1.7021
	2198	**0.0000**	**2198**	**0.0000**	**2198**	0.0045	**2198**	0.1365	**2198**	0.1683
	1805	**0.0000**	**1805**	**0.0000**	**1805**	**0.0000**	**1805**	**0.0000**	**1805**	**0.0000**
	2266	**0.0000**	**2266**	**0.0000**	**2266**	**0.0000**	**2266**	0.1721	**2266**	0.1765
	2056	**0.0000**	**2056**	**0.0000**	**2056**	**0.0000**	**2056**	0.0049	**2056**	0.0632
	1293	**0.0000**	**1293**	0.0232	**1293**	**0.0000**	**1293**	0.2320	**1293**	0.1469
	1933	**0.0000**	**1933**	**0.0000**	**1933**	0.0052	**1933**	0.3001	**1933**	0.4190
	2064	**0.0000**	**2064**	**0.0000**	**2064**	**0.0000**	**2064**	0.0581	**2064**	0.4457
	2093	**0.0000**	**2093**	0.0096	**2093**	**0.0000**	**2093**	0.0334	**2093**	0.0956
60 × 10	**2548**	**0.0078**	**2548**	0.1138	2550	0.2198	2556	0.9027	2569	1.4717
	2312	**0.0087**	**2312**	0.0173	**2312**	0.0216	**2312**	0.1125	2314	0.1557
	2493	**0.0000**	**2493**	**0.0000**	**2493**	**0.0000**	**2493**	**0.0000**	**2493**	**0.0000**
	1318	1.0687	**1310**	**0.8779**	1323	1.7634	1354	4.6947	1390	7.7481
	2294	**0.0000**	**2294**	0.1351	**2294**	0.2180	2312	2.5327	2372	4.5205
	2303	**0.0043**	**2303**	0.0261	**2303**	0.0261	**2303**	0.8033	**2303**	1.3200
	2275	0.0132	**2275**	**0.0000**	**2275**	0.2022	2286	1.0022	2296	1.4374
	2807	**0.0000**	**2807**	0.0143	**2807**	**0.0000**	**2807**	0.1603	**2807**	0.1176
	2091	**0.0000**	**2091**	**0.0000**	**2091**	**0.0000**	**2091**	0.0191	**2091**	0.0670
	2314	**0.0000**	**2314**	**0.0000**	**2314**	**0.0000**	**2314**	0.4322	2316	0.6698

Table 2. *Cont.*

Instance $m \times n$	HFA-VND		MBO		GATS		GAVNS		GA	
	C_{max}	RPI (%)	C_{max}	RPI (%)	C_{max}	RPI (%)	C_{max}	RPI (%)	C_{max}	RPI (%)
80 × 5	**3032**	**0.0000**	**3032**	**0.0000**	**3032**	**0.0000**	**3032**	0.1682	**3032**	0.1748
	2743	**0.0000**	**2743**	0.0292	**2743**	0.0146	**2743**	0.1568	2746	0.1969
	2963	**0.0000**	**2963**	**0.0000**	**2963**	**0.0000**	**2963**	0.0540	**2963**	0.2565
	1028	**0.3307**	1030	0.5545	1031	0.6712	1038	1.8580	1045	2.9961
	1466	**0.0205**	**1466**	0.1296	**1466**	0.1501	1467	0.7503	1472	0.9891
	2953	**0.0000**	**2953**	**0.0000**	**2953**	**0.0000**	**2953**	**0.0000**	**2953**	0.0847
	3049	**0.0000**	**3049**	**0.0000**	**3049**	**0.0000**	**3049**	0.0656	**3049**	0.0918
	1029	**0.1069**	1031	0.4859	**1029**	0.5831	1035	1.4383	1036	2.3032
	2933	**0.0000**	**2933**	**0.0000**	**2933**	**0.0000**	**2933**	0.0034	**2933**	0.0136
	3411	**0.0000**	**3411**	**0.0000**	**3411**	**0.0000**	**3411**	0.0967	**3411**	0.0850
80 × 10	**3330**	**0.0210**	**3330**	0.0631	**3330**	0.0300	3336	0.4174	3340	0.7838
	3001	**0.0000**	**3001**	0.0233	**3001**	0.0133	3009	0.6698	3031	1.6028
	3093	**0.0000**	**3093**	0.4753	**3093**	**0.0000**	**3093**	0.1843	3094	0.6790
	3254	**0.0000**	**3254**	**0.0000**	**3254**	0.0277	**3254**	0.3534	3263	0.6669
	3335	**0.0000**	**3335**	0.0090	**3335**	0.0300	**3335**	0.5607	3339	0.8126
	3390	**0.0000**	**3390**	**0.0000**	**3390**	**0.0000**	**3390**	0.1917	**3390**	0.5575
	3220	**0.0000**	**3220**	**0.0000**	**3220**	**0.0000**	**3220**	0.2143	3221	0.2640
	3318	**0.0000**	**3318**	0.0060	**3318**	**0.0000**	3320	0.3647	3321	0.7022
	3720	**0.0000**	**3720**	0.0269	**3720**	0.0054	**3722**	0.2124	3723	0.4086
	3235	**0.0000**	**3235**	**0.0000**	**3235**	**0.0000**	**3235**	0.0587	**3235**	0.1731
100 × 5	**1529**	**0.0131**	**1529**	0.2158	**1529**	0.1504	1535	0.7194	1540	1.1969
	3646	**0.0000**	**3646**	**0.0000**	**3646**	**0.0000**	**3646**	0.0439	**3646**	0.1015
	4494	**0.0000**	**4494**	**0.0000**	**4494**	**0.0000**	**4494**	0.0490	4495	0.1313
	3826	**0.0000**	**3826**	**0.0000**	**3826**	**0.0000**	**3826**	0.0261	**3826**	0.0497
	1269	**0.0709**	1270	0.1970	1271	0.5201	1272	1.8991	1289	2.6084
	3906	**0.0000**	**3906**	**0.0000**	**3906**	0.0410	**3906**	0.0205	**3906**	0.2125
	1920	**0.0417**	**1920**	0.1823	1921	0.1563	1922	0.4479	1925	0.6510
	3744	**0.0027**	**3744**	0.0053	**3744**	0.0374	3747	0.2404	3751	0.4006
	3453	**0.0000**	**3453**	0.0174	**3453**	**0.0000**	**3453**	0.0203	**3453**	0.4721
	851	**0.9220**	**846**	1.3475	856	1.9385	876	4.7518	889	7.1040
100 × 10	**4017**	**0.0149**	**4017**	0.0797	**4017**	0.0697	4037	1.0456	4081	2.3973
	4157	**0.0168**	**4157**	0.0433	**4157**	0.1203	4181	1.2846	4251	3.2090
	3854	**0.0130**	**3854**	0.1349	3855	0.2206	3866	0.4229	3875	0.9471
	3687	**0.0000**	**3687**	0.0434	**3687**	0.0108	**3687**	0.6618	3704	1.6273
	3931	**0.0000**	**3931**	0.0153	**3931**	**0.0000**	**3931**	0.3104	3942	1.0150
	3457	**0.0000**	**3457**	0.0579	**3457**	0.0289	**3457**	0.4484	3473	0.8360
	3671	**0.0054**	**3671**	0.0082	**3671**	0.1171	3683	0.5802	3689	1.0488
	4080	**0.0123**	**4080**	0.0368	**4080**	0.0735	**4080**	0.1642	4085	0.4240
	3930	**0.0000**	**3930**	**0.0000**	**3930**	**0.0000**	**3930**	0.3791	3933	0.5649
	4318	**0.0903**	4320	0.1065	4320	0.1667	4326	0.6901	4346	1.6350
Numbers of best C_{max} or RPI	**97**	**98**	89	38	88	51	66	8	44	4
Averages of RPI	**0.0005**		0.0017		0.0013		0.0069		0.0132	
Variances of RPI	**0.000004**		0.000009		0.000011		0.000106		0.000307	

In the Table 2, the bolded parts represent the best results obtained from these different algorithms under the corresponding instances. The comparison test results and the statistical information for different algorithms in Table 2 show that HFA-VND can achieve 97 best C_{max} and 98 best averages of *RPI* from the 100 instances. On the contrary, MBO, GATS, GAVNS, and GA cannot achieve the best results under dozens of instances. Moreover, the averages of *RPI* for each instance obtained from the other algorithms are inferior to those solved from HFA-VND. Through the further analysis for the test results in Table 2, GATS and MBO, which are often designed for complex scheduling problems, are superior to the common optimization algorithms (GAVNS and GA), and the hybrid algorithms (MBO, GATS, and GAVNS) are superior to the single algorithm (GA). Seen from the boxplot shown in Figure 6, the test results obtained from HFA-VND have more concentrated data distribution and fewer outliers. Therefore, it can be proved that HFA-VND proposed in this work is effective and stable for solving SCPPLSP under the multiple dynamic events and is superior to the current advanced optimization algorithms in the field of workshop scheduling.

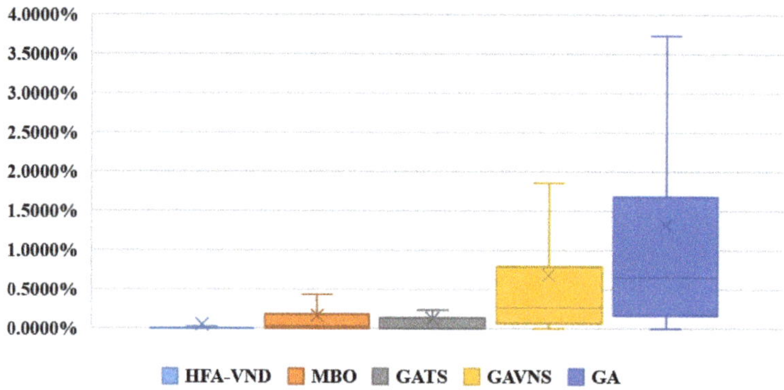

Figure 6. The boxplot for the averages of *RPI* obtained from different algorithms under each instance.

5.3. Case Study

In this section, the actual processing parameters and order data from a typical semiconductor chip packaging enterprise are used for a case test. In this enterprise, the semiconductor chip packaging production line contains 7 main processing stages, namely the grinding and slicing, attaching, wire bonding, molding, curing and plating, Trim/Form and marking, and testing, as shown in Figure 7. In the packaging production line, various types of wafers are packaged to various types of semiconductor chips. In the production workshop of the enterprise, each processing stage contains several packaging equipment or processing units.

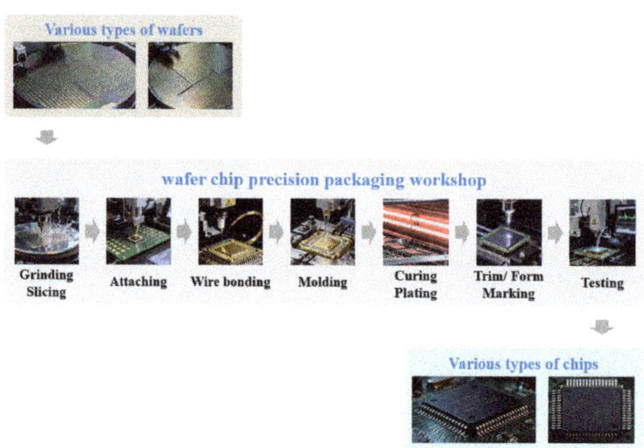

Figure 7. The semiconductor chip packaging processing flow.

Here, we use an actual order of this enterprise, which includes 60 types of chips, for a case test. In the test, the three types of dynamic events are simulated to occur randomly during the processing. To verify the effectiveness and superiority of the proposed method, MBO, GATS, GAVNS, and GA are introduced again for comparative testing. Moreover, the scheduling rules LPT and LPTF, which are currently being used by the enterprise, are also added to the comparative testing. The algorithm settings are the same as those in Section 5.2. The comparison test results are recorded in Table 3, and the bolded part represents the best result obtained from these different algorithms under the case test. The test results of actual case in Table 3 show that the HFA-VND can achieve the best scheduling result. As the same as Section 5.2, MBO is also superior to the other algorithm for actual case, and hybrid algorithms also perform better than the single algorithm. Comparing the scheduling rules LPT and LPTF, intelligent optimization algorithm can obtain the more optimal results. It can be seen that the makespan obtained from the HFA-VND can decrease by about 7% compared to LPT and LPTF. The case tests show the HFA-VND is effective for the SCPPLSP in the actual environment and can improve the production efficiency for semiconductor chips' packaging enterprise.

Table 3. The comparison test results for actual case.

HFA-VND	MBO	GATS	GAVNS	GA	LPT	LPTF
C_{max} (min)	C_{max} (min)	C_{max} (min)	C_{max} (min)	C_{max} (min)	C_{max} (min)	C_{max} (min)
1513	1524	1536	1538	1574	1611	1618

6. Conclusions

This paper proposed an HFA-VND to solve the SCPPLSP under multiple dynamic events. Firstly, the SCPPLSP is described in detail. After analysis on the three types of dynamic events, namely machine breakdown, emergency order inserting, and original order modification, the rescheduling mode based on rolling horizon technology is adopted. Based on the problem characteristics, the MILP model for SCPPLSP is established. Then, inspired by the design idea of standard FA, an improved FA framework is constructed. In order to efficiently describe the individuals, an encoding mechanism based on the sequence of chip types is used. Correspondingly, to save on the waiting times of sub-lots and the idle times of machines, the decoding mechanism based on two heuristic rules is presented to obtain the rescheduling plan. To ensure the diversity and superiority of the initial individuals, the random individuals and the NEH rule based on LPT and LPTF are applied to generating the initial population. In the process of population evolution, the individual movement strategy based on PBX improves the solutions quality, and the random perturbation based on the swap operator and insertion operator is introduced to enhance search capability. The local search mechanism based on VND for the optimal individuals ensures the sufficient exploration in their neighborhood. At last, the comparative test results under the 100 instances generated randomly show that the HFA-VND is effective for SCPPLSP and is superior to the current advanced algorithms, namely MBO, GATS, GAVNS, and GA. Additionally, a case test from semiconductor chips' precision packaging enterprise verified that the HFA-VND is effective for the actual production, and can accelerate the production efficiency for the enterprise.

In future studies, more types of dynamic events will be introduced, such as fuzzy processing time, lack of raw materials, periodic preventive maintenance, and so on. In addition, more optimization objectives will be involved, like energy consumption, utilization rate of equipment, production cost and tardiness. Additionally, the further improvement of the optimization algorithms will still be one of the important research directions.

Author Contributions: Z.W.: conceptualization, methodology, formal analysis, writing—original draft preparation, writing—review and editing, visualization, supervision, project administration, funding acquisition. W.F.: investigation, resources, data curation. Y.Y.: software, validation. All authors have read and agreed to the published version of the manuscript.

Funding: The study is supported by the National Natural Science Foundation of China under Grant number 52205522.

Data Availability Statement: The data that support the findings of this study are available from the corresponding author, upon reasonable request.

Conflicts of Interest: The authors declare that they have no competing interests.

Abbreviations

Parameters:

m	number of the processing stages
n	number of the chip types
k	index of processing stage, $k\in\{1,2,\ldots,m\}$
i	index of chip type, $i\in\{1,2,\ldots,n\}$
j	index of processing unit
e	index of sub-lot
$\boldsymbol{M}$	set of all processing stages
$\boldsymbol{J}$	set of all chip types
$\boldsymbol{M_k}$	set of all processing units at processing stage k
m_k	number of the processing units at processing stage k
l_i	number of sub-lots for chip type i
$tP_{i,k}$	processing time of sub-lot for chip type i at processing stage k
$tS_{i,e,k}$	starting time of sub-lot e for the chip type i at processing stage k
$tC_{i,e,k}$	completion time of sub-lot e for chip type i at processing stage k
tC_i	completion time for chip type i at processing stage m
tB	the time when the processing unit breaks down
tR	the time when the faulty processing unit is restored
$O_{i,e,k}$	operation of sub-lot e for chip type i at processing stage k
***State* 1**	State set 1
***State* 2**	State set 2
***State* 3**	State set 3
$bD_{i,j,k}$	1 if chip type i is on machine j at processing stage k, 0 otherwise
$bS_{i,i',k}$	1 if chip type i is superior to type i' on same machine at stage k, 0 otherwise
Q	a sufficiently large positive integer

Decision variables:

$tS'_{i,e,k}$	starting time of sub-lot e for chip type i at processing stage k in rescheduling
$tC'_{i,e,k}$	completion time of sub-lot e for chip type i at processing stage k in rescheduling
$bD'_{i,j,k}$	1 if chip type i is on machine j at processing stage k in rescheduling, 0 otherwise
$bS'_{i,i',k}$	1 if chip type i is superior to type i' on same machine at stage k, 0 otherwise

References

1. Zhou, D.; Gong, S.; Wang, Z.; Li, D.; Lu, H. Error analysis based on error transfer theory and compensation strategy for LED chi visual localization systems. *J. Intell. Manuf.* **2021**, *32*, 1345–1359. [CrossRef]
2. Azzoni, P.; di Paola Galloni, J.L.; Serpanos, D. Chips shortage and chips acts: The inside view. *Computer* **2023**, *56*, 113–11 [CrossRef]
3. Wang, Z.; Zhou, D.; Gong, S. Uncalibrated visual positioning using adaptive Kalman Filter with dual rate structure for wafer chi in LED packaging. *Measurement* **2022**, *191*, 110829. [CrossRef]

4. Tang, H.; Huang, J.; Ren, C.; Shao, Y.; Lu, J. Integrated scheduling of multi-objective lot-streaming hybrid flowshop with AGV based on deep reinforcement learning. *Int. J. Prod. Res.* **2025**, *63*, 1275–1303. [CrossRef]
5. Wang, G.; Ren, M. Recall cost-time tradeoffs for remanufacturing shop lot streaming scheduling problem with non mixed production using an improved non-dominated sorting genetic algorithm. *Int. J. Ind. Eng. Comput.* **2025**, *16*, 1055–1076. [CrossRef]
6. Tsubone, H.; Ohba, M.; Uetake, T. The impact of lot sizing and sequencing on manufacturing performance in a two-stage hybrid flow shop. *Int. J. Prod. Res.* **1996**, *34*, 3037–3053. [CrossRef]
7. Shiue, Y.R.; Lee, K.C.; Su, C.T. Development of dynamic scheduling in semiconductor manufacturing using a Q-learning approach. *Int. J. Comput. Integr. Manuf.* **2022**, *35*, 1188–1204. [CrossRef]
8. Holguin Jimenez, S.; Trabelsi, W.; Sauvey, C. Multi-objective optimisation for permutation flow shop rescheduling problems with mixed blocking constraints. *Int. J. Syst. Sci.* **2025**, *12*, 2541379. [CrossRef]
9. Meng, L.; Zhang, C.; Zhang, B.; Gao, K.; Ren, Y.; Sang, H. MILP modeling and optimization of multi-objective flexible job shop scheduling problem with controllable processing times. *Swarm Evol. Comput.* **2023**, *82*, 101374. [CrossRef]
10. Zhang, B.; Pan, Q.K.; Gao, L.; Zhang, X.L.; Peng, K.K. A multi-objective migrating birds optimization algorithm for the hybrid flowshop rescheduling problem. *Soft Comput.* **2019**, *23*, 8101–8129. [CrossRef]
11. Yang, X.S. Firefly algorithms for multimodal optimization. *Int. Symp. Stoch. Algorithms* **2009**, *5792*, 169–178.
12. Fan, B.; Yang, W.; Zhang, Z. Solving the two-stage hybrid flow shop scheduling problem based on mutant firefly algorithm. *J. Ambient Intell. Humaniz. Comput.* **2019**, *10*, 979–990. [CrossRef]
13. Wang, Z.; Shen, L.; Li, X.; Gao, L. An improved multi-objective firefly algorithm for energy-efficient hybrid flowshop rescheduling problem. *J. Clean Prod.* **2023**, *385*, 135738. [CrossRef]
14. Xu, E.B.; Zou, F.F.; Shan, P.P.; Wang, Z.Y.; Shi, B.X. An improved multi-objective firefly algorithm for integrated scheduling approach in manufacturing and assembly considering time-sharing step tariff. *Adv. Prod. Eng. Manag.* **2024**, *19*, 5–20. [CrossRef]
15. Chen, D.; Zhang, J.; Wu, L.; Zhang, P.; Lv, Y.; Wang, J.; Wang, H. Optimisation method for semiconductor wafer manufacturing system scheduling: Reinforcement learning with decision graph guiding. *J. Manuf. Syst.* **2025**, *82*, 1158–1170. [CrossRef]
16. Immordino, A.; Stöckermann, P.; Hayen, N.; Altenmüller, T.; Susto, G.A.; Gebser, M.; Schekotihin, K.; Seidel, G. Explainable AI for reinforcement learning based dynamic scheduling solutions in semiconductor manufacturing. *J. Intell. Manuf.* **2025**, 1–17. [CrossRef]
17. Yang, F.; Wu, N.; Qiao, Y.; Zhou, M.; Li, Z. Scheduling of single-arm cluster tools for an atomic layer deposition process with residency time constraints. *IEEE Trans. Syst. Man Cybern.-Syst.* **2016**, *47*, 502–516. [CrossRef]
18. Wang, J.; Gao, P.; Zheng, P.; Zhang, J.; Ip, W.H. A fuzzy hierarchical reinforcement learning based scheduling method for semiconductor wafer manufacturing systems. *J. Manuf. Syst.* **2021**, *61*, 239–248. [CrossRef]
19. Mönch, L.; Fowler, J.W.; Dauzère-Pérès, S.; Mason, S.J.; Rose, O. A survey of problems, solution techniques, and future challenges in scheduling semiconductor manufacturing operations. *J. Sched.* **2011**, *14*, 583–599. [CrossRef]
20. Siebert, M.; Bartlett, K.; Kim, H.; Ahmed, S.; Lee, J.; Nazzal, D.; Nemhauser, G.; Sokol, J. Lot targeting and lot dispatching decision policies for semiconductor manufacturing: Optimisation under uncertainty with simulation validation. *Int. J. Prod. Res.* **2018**, *56*, 629–641. [CrossRef]
21. Calmels, D. An iterated local search procedure for the job sequencing and tool switching problem with non-identical parallel machines. *Eur. J. Oper. Res.* **2022**, *297*, 66–85. [CrossRef]
22. Chen, D.; Zhang, J.; Wu, L.; Zhang, P.; Wang, M. Industrial data space application framework for semiconductor wafer manufacturing system scheduling. *J. Manuf. Syst.* **2024**, *77*, 464–482. [CrossRef]
23. Hsu, H.T.; Chang, S.C. Effective deep reinforcement learning for dynamic machine allocation: A case study on metal sputtering tools. *IEEE Trans. Semicond. Manuf.* **2025**, *38*, 430–438. [CrossRef]
24. Tao, F.; Qi, Q. Make more digital twins. *Nature* **2019**, *573*, 490–491. [CrossRef] [PubMed]
25. Ding, L.; Luo, D.; Mudassar, R.; Yue, L.; Meng, L. A novel deep self-learning method for flexible job-shop scheduling problems with multiplicity: Deep reinforcement learning assisted the fluid master-apprentice evolutionary algorithm. *Swarm Evol. Comput.* **2025**, *94*, 101907. [CrossRef]
26. Park, I.B.; Park, J. Scalable scheduling of semiconductor packaging facilities using deep reinforcement learning. *IEEE Trans. Cybern.* **2021**, *53*, 3518–3531. [CrossRef]
27. Lee, Y.H.; Lee, S. Deep reinforcement learning based scheduling within production plan in semiconductor fabrication. *Expert Syst. Appl.* **2022**, *191*, 116222. [CrossRef]
28. Tian, Z.; Jiang, X.; Liu, W.; Zhao, B.; Tan, Q.; Li, Z. Energy-efficient scheduling for lot-streaming flexible job-shop considering machine deterioration and maintenance. *J. Manuf. Syst.* **2025**, *82*, 927–946. [CrossRef]
29. Zhang, B.; Sang, H.; Meng, L.; Jiang, X.; Lu, C. Knowledge-and data-driven hybrid method for lot streaming scheduling in hybrid flowshop with dynamic order arrivals. *Comput. Oper. Res.* **2025**, *184*, 107244. [CrossRef]

30. Zhao, F.; Yin, F.; Zhang, J.; Xu, T.P. A learning-based co-evolution optimization framework for energy-aware distributed heterogeneous flexible flow shop lot-streaming scheduling problem. *Expert Syst. Appl.* **2026**, *296*, 128986. [CrossRef]
31. Yue, L.; Huang, J.; Ding, L.; Cai, X.; Sun, Y.; Zou, T. Integrated lot-sizing and scheduling for parallel mixed-model automotive production lines with transportation resource constraints. *J. Ind. Inf. Integr.* **2025**, *48*, 100945. [CrossRef]
32. Zhang, B.; Pan, Q.K.; Gao, L.; Zhang, X.L.; Sang, H.Y.; Li, J.Q. An effective modified migrating birds optimization for hybrid flowshop scheduling problem with lot streaming. *Appl. Soft. Comput.* **2017**, *52*, 14–27. [CrossRef]
33. Defersha, F.M.; Chen, M. Mathematical model and parallel genetic algorithm for hybrid flexible flowshop lot streaming problem. *Int. J. Adv. Manuf. Technol.* **2012**, *62*, 249–265. [CrossRef]
34. Li, J.Q.; Tao, X.R.; Jia, B.X.; Han, Y.Y.; Liu, C.; Duan, P.; Zhang, Z.X.; Sang, H.Y. Efficient multi-objective algorithm for the lot-streaming hybrid flowshop with variable sub-lots. *Swarm Evol. Comput.* **2020**, *52*, 100600. [CrossRef]
35. Wang, S.; Kurz, M.; Mason, S.J.; Rashidi, E. Two-stage hybrid flow shop batching and lot streaming with variable sublots and sequence-dependent setups. *Int. J. Prod. Res.* **2019**, *57*, 6893–6907. [CrossRef]
36. Nejati, M.; Mahdavi, I.; Hassanzadeh, R.; Mahdavi-Amiri, N.; Mojarad, M. Multi-job lot streaming to minimize the weighted completion time in a hybrid flow shop scheduling problem with work shift constraint. *J. Adv. Manuf. Technol.* **2014**, *70*, 501–514. [CrossRef]
37. Nejati, M.; Mahdavi, I.; Hassanzadeh, R.; Mahdavi-Amiri, N. Lot streaming in a two-stage assembly hybrid flow shop scheduling problem with a work shift constraint. *J. Ind. Prod. Eng.* **2016**, *33*, 459–471. [CrossRef]
38. Lalitha, J.L.; Mohan, N.; Pillai, V.M. Lot streaming in [N-1](1) + N(m) hybrid flow shop. *J. Manuf. Syst.* **2017**, *44*, 12–21. [CrossRef]
39. Novas, J.M. Production scheduling and lot streaming at flexible job-shops environments using constraint programming. *Comput. Ind. Eng.* **2019**, *136*, 252–264. [CrossRef]
40. Cheng, M.; Sarin, S.C.; Singh, S. Two-stage, single-lot, lot streaming problem for a 1 + 2 hybrid flow shop. *J. Glob. Optim.* **2016**, *66*, 263–290. [CrossRef]
41. Vivek, P.; Saravanan, R.; Chandrasekaran, M.; Pugazhenthi, R. Heuristic for critical machine based a lot streaming for two-stage hybrid production environment. *IOP Conf. Ser. Mater. Sci. Eng.* **2017**, *183*, 012030. [CrossRef]
42. Wang, R.; Cheung, C.F.; Zang, Y.; Wang, C.; Liu, C. Material removal rate optimization with bayesian optimized differential evolution based on deep learning in robotic polishing. *J. Manuf. Syst.* **2025**, *78*, 178–186. [CrossRef]
43. Qiao, F.; Ma, Y.; Zhou, M.; Wu, Q. A novel rescheduling method for dynamic semiconductor manufacturing systems. *IEEE Trans. Syst. Man Cybern.-Syst.* **2018**, *50*, 1679–1689. [CrossRef]
44. Wei, S.; Tang, H.; Li, X.; Lei, D.; Wang, X.V. An improved memetic algorithm for multi-objective resource-constrained flexible job shop inverse scheduling problem: An application for machining workshop. *J. Manuf. Syst.* **2024**, *74*, 264–290. [CrossRef]
45. Wang, D.J.; Liu, F.; Wang, Y.Z.; Jin, Y. A knowledge-based evolutionary proactive scheduling approach in the presence of machine breakdown and deterioration effect. *Knowl.-Based Syst.* **2015**, *90*, 70–80. [CrossRef]
46. Ding, L.; Guan, Z.; Luo, D.; Yue, L. Data-driven hierarchical multi-policy deep reinforcement learning framework for multi-objective multiplicity dynamic flexible job shop scheduling. *J. Manuf. Syst.* **2025**, *80*, 536–562. [CrossRef]
47. Yue, L.; Peng, K.; Ding, L.; Mumtaz, J.; Lin, L.; Zou, T. Two-stage double deep Q-network algorithm considering external non-dominant set for multi-objective dynamic flexible job shop scheduling problems. *Swarm Evol. Comput.* **2024**, *90*, 101660. [CrossRef]
48. He, X.; Dong, S.; Zhao, N. Research on rush order insertion rescheduling problem under hybrid flow shop based on NSGA-III. *Int. J. Prod. Res.* **2020**, *58*, 1161–1177. [CrossRef]
49. Mou, J.; Gao, L.; Guo, Q.; Mu, J. A hybrid heuristic algorithm for flowshop inverse scheduling problem under a dynamic environment. *Cluster Comput.* **2017**, *20*, 439–453. [CrossRef]
50. Mou, J.; Duan, P.; Gao, L.; Liu, X.; Li, J. An effective hybrid collaborative algorithm for energy-efficient distributed permutation flow-shop inverse scheduling. *Future Gener. Comp. Syst.* **2022**, *128*, 521–537. [CrossRef]
51. Zhang, Y.; Hu, X.; Cao, X.; Wu, C. An efficient hybrid integer and categorical particle swarm optimization algorithm for the multi-mode multi-project inverse scheduling problem in turbine assembly workshop. *Comput. Ind. Eng.* **2022**, *169*, 108148. [CrossRef]
52. Qiao, Y.; Wu, N.; Yang, F.; Zhou, M.; Zhu, Q.; Qu, T. Robust scheduling of time-constrained dual-arm cluster tools with wafer revisiting and activity time disturbance. *IEEE Trans. Syst. Man Cybern.-Syst.* **2017**, *49*, 1228–1240. [CrossRef]
53. Liu, J.; Qiao, F.; Kong, W. Scenario-based multi-objective robust scheduling for a semiconductor production line. *Int. J. Prod. Res.* **2019**, *57*, 6807–6826. [CrossRef]
54. Lee, M.; Morrison, J.R.; Kalir, A.A. Practical queueing models for preventive maintenance plan optimization: Multiple maintenance types and numerical studies. *IEEE Trans. Semicond. Manuf.* **2022**, *34*, 104–114. [CrossRef]

5. Li, X.; Gao, L. An effective hybrid genetic algorithm and tabu search for flexible job shop scheduling problem. *Int. J. Prod. Econ.* **2016**, *174*, 93–110. [CrossRef]

. Long, J.; Zheng, Z.; Gao, X. Dynamic scheduling in steelmaking-continuous casting production for continuous caster breakdown. *Int. J. Prod. Res.* **2017**, *55*, 3197–3216. [CrossRef]

micromachines

MDPI

Article

Investigation into Laser-Vibration-Assisted Cutting of Single-Crystal Silicon by Molecular Dynamics Simulation

Jianning Chu [1], Yichen Yang [2], Yikai Zang [3,*], Jinyang Ke [3], Ziyue Wang [4], Chen Chen [5], Jifei He [5], Aijiang Xu [5] and Zhongdi She [3,*]

1 Hubei Key Laboratory of Modern Manufacturing Quality Engineering, School of Mechanical Engineering, Hubei University of Technology, Wuhan 430068, China; jianning_chu@163.com
2 Marine Design and Research Institute of China, Shanghai 200011, China
3 State Key Laboratory of Intelligent Manufacturing Equipment and Technology, School of Mechanical Science and Engineering, Huazhong University of Science and Technology, Wuhan 430074, China
4 School of Intelligent Manufacturing, Jianghan University, Wuhan 430056, China
5 Jiangsu Nuclear Power Corporation, Lianyungang 222044, China
* Correspondence: zangyk@hust.edu.cn (Y.Z.); zhongdishe@hust.edu.cn (Z.S.)

Abstract: It is difficult to achieve ultra-precision machining (UPM) on semiconductor materials like single-crystal silicon because of their hardness and brittleness. To solve this issue, numerous field-assisted machining systems and their combinations have been suggested and developed. However, the difficulty in directly observing the physical variables limits our comprehension of the in-depth machining mechanisms of field-assisted machining. In this work, we investigated the machining mechanism of single-crystal silicon under the combination of laser heating and tool vibration using molecular dynamics (MD) simulations. The effect of tool vibration trajectory determined by different tool edge radii is discussed under the condition of raising temperature. The simulation results indicate that the surface morphology is closely related to vibration and heating parameters. Raising the cutting temperature causes a reversed relation between tool edge radius and surface roughness. While the subsurface damage and internal stress are also determined by the tool edge radius and cutting temperature. The findings in this simulation could help to improve the understanding of machining mechanics in multi-field-assisted machining.

Keywords: field-assisted machining; molecular dynamics simulation; single-crystal silicon; crystal defects; material removal mechanism

1. Introduction

Owing to the exceptional optical and electronic properties, single-crystal silicon (Si) is widely employed in Complementary Metal Oxide Semiconductor (CMOS) [1] and photovoltaics [2]. The performance of these devices depends on the machined surface quality of Si components, which necessitates the use of ultra-precision machining (UPM) technologies such as nano-cutting [3], nano-grinding [4], and nano-polishing [5]. However, the primary barriers to achieving high-quality surfaces on single-crystal Si are subsurface damage and unwanted surface defects and cutting tool wear caused by its brittleness and hardness [6,7].

Over the past few decades, various field-assisted machining technologies have been developed to improve the machinability of hard and brittle materials. Laser-assisted machining (LAM) emerged as lasers matured in the 1960s, heating the workpiece with a focused laser spot to thermally soften the workpiece material and drive the removal mode into a ductile regime, thereby reducing the cutting forces and tool wear [8]. In addition to

LAM, vibration-assisted machining has been applied in the manufacturing industry since the 1960s [9]. In 1994, Shamoto and Moriwaki proposed elliptical vibration machining (EVM) [10,11] to achieve high-quality surfaces with complex microstructures on brittle materials. By applying high-frequency elliptical vibration to the cutting tool, the cutting motion becomes intermittent, and the material loading is significantly reduced, which enables ductile-regime machining of brittle materials with lower cutting forces and less cutting heat than ordinary cutting [12]. Magnetic field-assisted machining (MFAM), which was first proposed by Yip et al. [13], has been verified as a promising method to achieve UPM of difficult-to-cut materials. Applying a magnetic field is advantageous for improving the plastic-deformation ability of magnetic-conductive materials such as Ti alloys [14]. In recent years, multi-field-assisted machining technologies have been developed to address the bottlenecks of single assistive fields in UPM. Xing et al. [15] combined the EVM with MFAM methods to improve the machining performance of high-entropy alloys. Their results demonstrated that multi-field-assisted machining combines the advantages of single-field machining and effectively enhances the machinability of high-entropy alloys. Ke et al. [16] combined ion implantation with EVM in the machining of sapphire. They used high-energy ions to modify the workpiece surface. Combined with tool vibration, a great enhancement in the machinability of sapphire was achieved. Zhang et al. [17] coupled LAM and EVM to investigate the fabrication process of silicon microlens arrays. Their results suggested that a combination of the single assistive fields broadened the applications of field-assisted machining. In multi-field-assisted machining, the material deformation behavior is more complex than ordinary machining and single-field-assisted machining, as the coupling effect of the assistive fields involves sophisticated interactions between workpieces and tools. Unfortunately, there are few investigations that focus on the machining mechanism during multi-field-assisted machining. And the coupled effect of multiple fields on material removal features and subsurface damage evolution is still unclear.

Based on Newton's equations of motion, molecular dynamics (MD) simulation has become a viable way to investigate the deformation mechanism of materials in UPM. It has been successfully applied in investigations of LAM [18], EVM [19], and ion-implantation-assisted machining [20]. Dai et al. [21] studied the machining mechanism of single-crystal Si in LAM by employing an intense laser beam to heat the workpiece material. Liu et al. [22] established an MD model to investigate the machining mechanism of reaction-bonded silicon carbide during LAM. They used a uniform initial temperature distribution to explore the effect of laser heating on material deformation and verified the simulation results with experiments. While for EVM, previous simulation results demonstrated that compared to ordinary machining, smaller internal stresses are introduced in the workpiece during machining, which leads to less subsurface damage [23]. Recently, to narrow the gap in parameters between MD simulation and cutting experiments, Liu et al. [24] proposed a new cutting model to simulate the transient cutting process in EVM. They suggested that the dominant removal mechanism of EVM experiences the transition from extrusion to shear in one vibration cycle. Based on this model, they explored the cutting mechanism during the multi-field-assisted cutting that combined LAM and EVM [25]. They analyzed the effect of cutting temperature on the material removal and subsurface damage formation, while the effect of tool vibration trajectory on the machining mechanism under different temperatures was not mentioned. In laser-vibration-assisted machining, tool vibration determines the material removal behavior in a different manner than EVM, since the properties of the workpiece material are significantly influenced by the rising temperature.

Therefore, in the present work, we used MD simulation to investigate the machining mechanism during laser-vibration-assisted machining of single-crystal Si. The material removal behavior and subsurface damage were discussed under different cutting temperatures and tool edge radii, which are important factors that determine the trajectory of the contact point between the tool and the workpiece. The results in this simulation could help to improve the understanding of machining mechanics in multi-field-assisted machining and provide a reference for parameter optimization in the machining process.

2. Simulation Method

In EVM, the cutting tool moves along the nominal cutting direction and vibrates in the nominal cutting direction and depth of cut direction, as shown in Figure 1. The tool displacement can be calculated by the following:

$$x(t) = A_x \sin(2\pi f t) - vt \quad (1)$$

$$z(t) = A_z \sin(2\pi f t + \varphi) \quad (2)$$

where A_x and A_z are the vibration amplitudes in the nominal cutting direction and depth of cut direction, respectively. While f, v, φ, and t are the vibration frequency, nominal cutting speed, phase shift, and time, respectively.

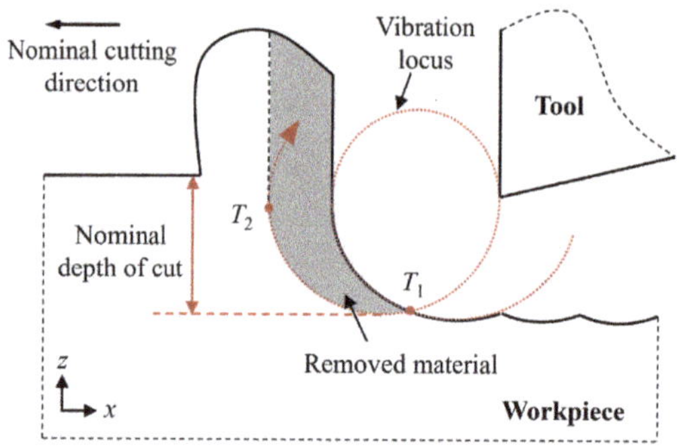

Figure 1. Schematic illustration of EVM. T_1 and T_2 represent the beginning and ending points of the contact stage in one vibration cycle, respectively. Reproduced with permission from C. Liu et al., *Nanoscale Res. Lett.*; published by *Springer Nature*, 2021 [25].

The simulation and data processing were performed by LAMMPS [26] and OVITO [27]. Figure 2 displays the adopted cutting model. The single-crystal Si workpiece was shaped according to the trajectory of the tool-workpiece contact point, and the edge of the workpiece was cut to reduce the computational cost [24]. To minimize the computational timesteps, only the contact stage in a vibration cycle (from T_1 to T_4) was simulated. The diamond tool was set as a rigid body while the workpiece was deformable. In accordance with the traditional MD paradigm for UPM, the workpiece atoms are categorized into Newton atoms, thermostat atoms, and boundary atoms [28,29]. The cutting simulation was conducted under different workpiece temperatures and tool edge radii. In previous research, the size of the thermally influenced region in the laser-assisted region is much larger than the deformed region in UPM, which indicates that the variation in temperature in the deformed region is quite small [22]. Therefore, in this research, a unified initial temperature condition was used to simulate the laser heating effect. The atomic interaction between Si-Si and C-C was characterized by ABOP [30], while the interaction between Si-C is described by the Morse potential, which has been shown to be a reliable option with desired calculation efficiency [31,32]. The Morse potential function can be expressed as follows:

$$U_{\text{Si-C}}(r_{ij}) = D_M\left[e^{-2a(r_{ij} - r_M)} - 2e^{-a(r_{ij} - r_M)}\right] \quad (3)$$

where D_M represents the cohesion energy, a is the modulus of elasticity, and r_M is the equilibrium distance between the atoms. The parameters of the Morse potential for Si-C are set as follows: D_M = 0.435 eV, a = 46.487 nm^{-1}, r_M = 0.19475 nm [33]. Table 1 lists detailed simulation parameters.

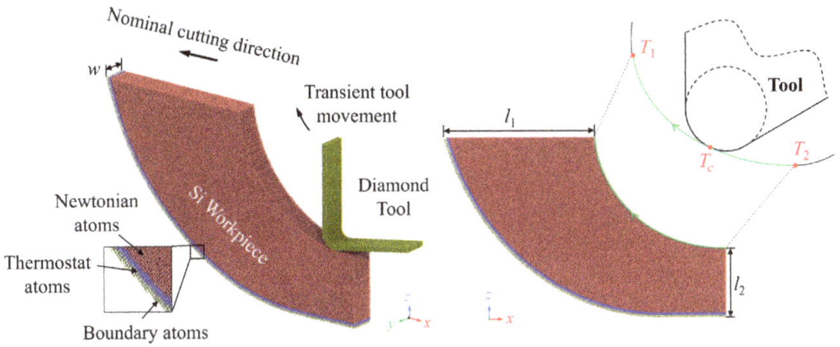

Figure 2. Illustration of the adopted MD cutting model. Reproduced with permission from C. Liu et al., *Nanoscale Res. Lett.*; published by *Springer Nature*, 2021 [25].

Table 1. Parameters of the MD model.

Parameters	Value
Size of workpiece ($l_1/l_2/w$)	50 nm/15 nm/8.7 nm
Crystal orientation	x:(001)[100]; y:(001)[010] (workpiece and tool)
Total atoms	Approximately 1.9–2.1 million
Cutting temperature	300 K, 600 K, 900 K, 1200 K
Vibration frequency (f)	500 MHz
Phase difference (φ)	90°
Vibration amplitude (A_x/A_d)	40 nm/40 nm
Nominal depth of cut	40 nm
Tool edge radius	5 nm, 10 nm, 15 nm, 20 nm, 25 nm
Nominal cutting speed	3 m/s
Timestep	1 fs

3. Results and Discussion

3.1. Surface Morphology

For single-crystal Si, the machined surface could derive from the theoretical surface due to the surface swelling effect, which is attributed to the strain energy release and phase transition of the metastable phases into the amorphous phase. After machining, the deformed workpiece material tends to return to its balanced position to release the strain energy, and the metastable phases could transform into an amorphous phase with lower density, which could cause obvious expansion of the material. Since the contact state and atomic flow are determined by the material removal thickness and cutting direction, the features of surface swelling in EVM can be significantly different from ordinary machining due to the constant tool vibration. Figure 3a shows a snapshot of the workpiece cut at 300 K by a tool with a 5 nm edge radius. In one vibration cycle, the surface swelling (atoms above the theoretical surface) gradually becomes obvious after the initial contact due to the rise in material removal thickness. As the cutting tool advances, the increase in upward motion of the cutting tool could release the compression and promote the atomic flow of workpiece material. Therefore, surface swelling becomes less obvious at the end of the contact stage.

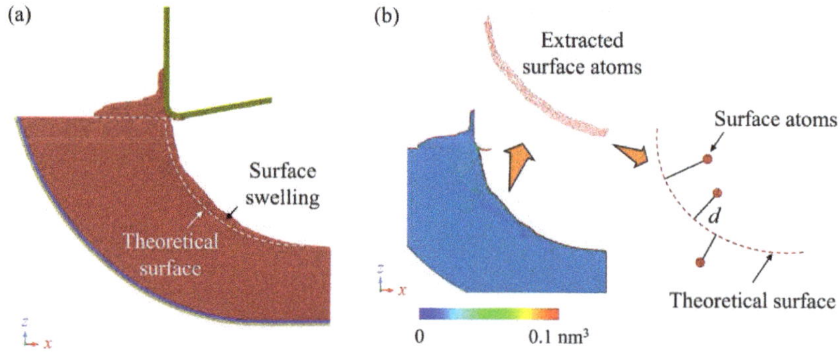

Figure 3. Morphology of the workpiece after cutting. (**a**) Swelling on the machined surface. (**b**) Estimation of surface roughness determined by positions of surface atoms, where the color bar indicates the atomic volume calculated by the Voronoi algorithm.

To evaluate the surface morphology after cutting, the machined surface roughness was calculated based on the position of surface atoms. In MD simulation, the atomic volume determined by the Voronoi algorithm can be used to identify the surface atoms. When atoms are positioned as the cores of a Voronoi cell, the volume of surface atoms in the simulation box can be significantly greater than that of atoms inside the workpiece [34], as shown in Figure 3b. Then, the machined surface roughness (Sa) can be estimated by the following:

$$\text{Sa} = \frac{1}{A}\iint |z(x, y)|dxdy \approx \frac{1}{n}\sum|d| \tag{4}$$

where d is the distance between the position of surface atoms and their theoretical locations, and n is the number of surface atoms. Figure 4a shows the calculated machined surface roughness after simulation under different conditions. It is observed that the surface roughness increases with rising tool edge radius when cutting at 300 K and 600 K, while a reversed trend is observed when the temperature increases to 900 K and 1200 K. The remarkably high surface roughness at 900 K and 5 nm tool edge radius is caused by the fracture of the workpiece. During the EVM process (shown in Figure 4b), the transient depth of cut and effective tool rake angle decrease as the tool edge radius increases, causing more surface plowing/rubbing and extrusion removal of workpiece material [24]. Therefore, surface swelling can be more obvious, which tends to increase the surface roughness. When the cutting temperature increases, the plastic deformability of the workpiece is enhanced, and the atomic flow of the surface disordered atoms can be promoted. When a cutting tool with a large edge is used, the tool-workpiece contact area is enlarged, and the surface undulation can be smoothed under high cutting temperature, causing a decrease in surface roughness.

3.2. Material Removal Behavior

Depending on the cutting conditions, the workpiece material can be removed by extrusion or shear during nanoscale cutting of single-crystal Si [35]. During the EVM process, the material removal mechanism varies continuously due to the continuous changes in the cutting direction and transient material removal thickness. The removal mechanism could transition from extrusion to shear during the workpiece-tool contact stage in a vibration cycle [24]. In MD simulation, the material removal behavior can be identified by analyzing the atomic displacement mapping. When elastic deformation occurs, atomic displacement grows gradually for atoms and their neighbors. An obvious interruption in the displacement magnitude indicates plastic deformation due to the destruction of the crystal structure.

Figure 5 shows the relative atomic displacement of different cutting stages in one vibration cycle, which is calculated from a period of 25 ps. In the initial contact stage, nearly no material is removed, and the main workpiece deformation features are surface abrasion and elastic deformation near the contact region, as shown in Figure 5a. As the material removal thickness increases, as shown in Figure 5b, extrusion of the disordered material merges as the dominant removal mechanism, and the disordered atoms are compressed into the subsurface workpiece, forming disordered clusters in the subsurface workpiece. With the advance of the cutting tool, the dislocation path can be observed owing to the increase in transient removal thickness and variation in cutting direction, as shown in Figure 5c. The promoted dislocation propagation and slip motion indicate that the dominant removal mechanism has switched from extrusion to a shear process. The remaining increase in atomic displacement on the machined surface indicates residual strain of the workpiece material. As the slip band extends to the uncut surface, the material is detached from the workpiece, as shown in Figure 5d.

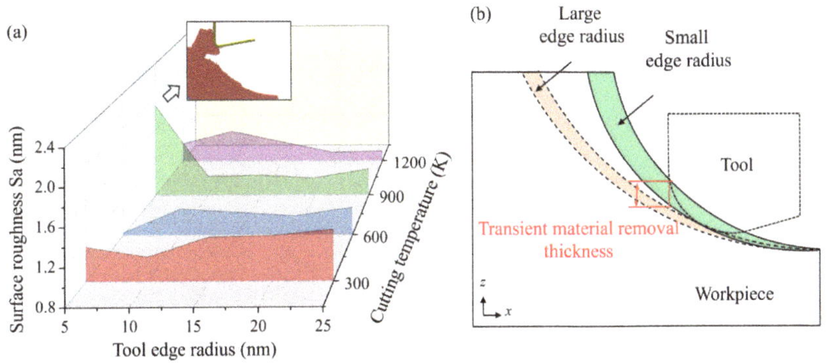

Figure 4. (**a**) The calculated surface roughness under different tool edge radii and cutting temperatures. (**b**) Illustration of the effect of tool edge radius on tool motion.

Figure 6 shows the atomic displacement mapping after cutting with tools with different edge radii and cutting temperatures, where the displacement was calculated from the initial position. Referring to Figure 5, an increase in the atomic displacement of the surface atoms indicates surface abrasion, while the slip motion suggests the shear process. When tools with larger edge radii are used, surface abrasion and extrusion removal are promoted since the contact area is enlarged with decreased transient removal thickness, as shown in Figure 6a–d. The plastic deformation mainly concentrates near the surface region, as inapparent atomic displacement was observed in the subsurface workpiece. When the cutting temperature increases, plastic deformation of the subsurface workpiece is promoted as more interruption in the atomic displacement distribution is observed, as shown in Figure 6e–h. In the surface abrasion and extrusion stage, obvious plastic deformation patterns are observed in the subsurface region owing to the enhanced plastic deformability at high temperatures. While in the shear stage, more slip bands are observed with voids formed in the deformed region since rolling and fracture of crystal grains driven by the pulling-up motion from the cutting tool are enhanced at high temperatures [25]. As shown in Figure 6h, the voids generated at high temperature have a minor influence on the quality of the final machined surface since most defects generated in the shear stage can be removed by following the cutting cycle during the EVM process.

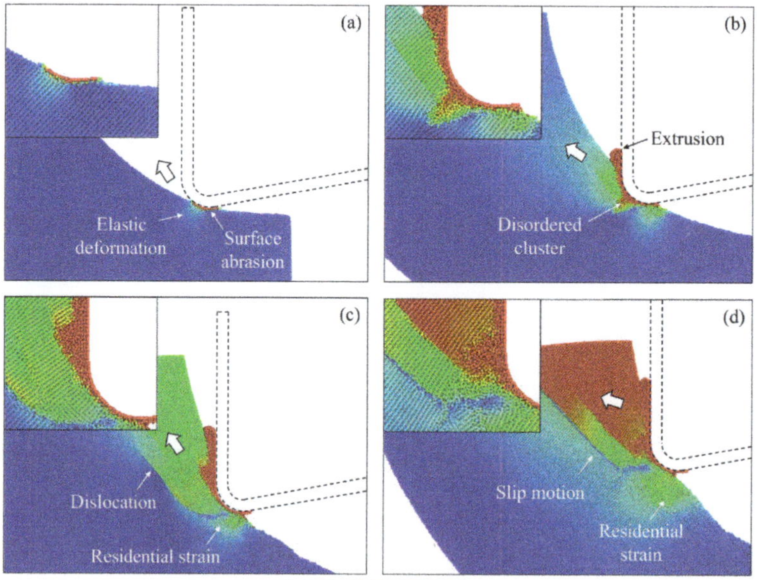

Figure 5. The atomic displacement mapping of workpiece atoms during different cutting stages, where the displacement was calculated over a period of 25 ps. (**a**) 100 ps. (**b**) 200 ps. (**c**) 300 ps. (**d**) 400 ps.

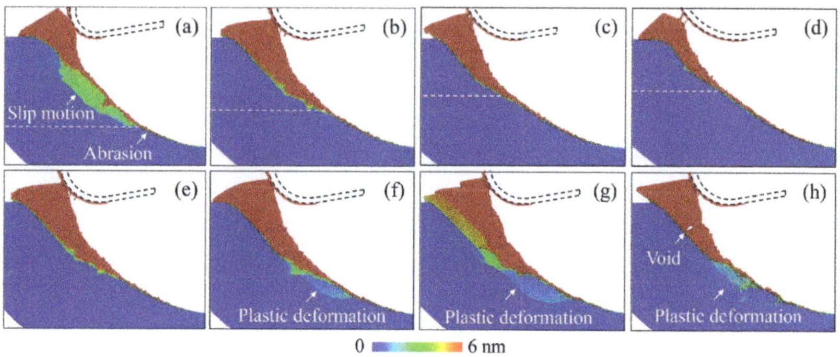

Figure 6. The atomic displacement mapping of workpiece atoms after cutting under different conditions, where the displacement was calculated from the initial position. (**a–d**) Cutting with tool edge radii of 10 nm, 15 nm, 20 nm, and 25 nm at 300 K, where the dashed line indicates the approximate position for the beginning of the slip motion. (**e–h**) Cutting at 300 K, 600 K, 900 K, and 1200 K with a tool edge radius of 15 nm.

3.3. *Crystal Defects*

The crystal structure of workpiece atoms at various cutting stages from the simulation at 300 K with a tool edge radius of 5 nm is depicted in Figure 7, where the crystal structure was determined using OVITO's common neighbor analysis (CNA) [36]. In the initial contact stage, the surface abrasion causes the generation of disordered atoms on the contact surface, as shown in Figure 7a. With the increase in material removal thickness, the high-pressure phase transition of the cubic diamond structure to other structures (including metastable phases and amorphous phases) emerges as a significant deformation process. These disordered atoms are extruded into chips ahead of the tool edge, as shown in Figure 7b. Meanwhile, an amorphous layer is left on the machined surface as a result of the surface abrasion and the transition of high-pressure phases into amorphous phases after machining. As the cutting tool further advances (see Figure 7c), the cutting direction changes and material removal thickness increases, facilitating the slip motion along the [111] direction. The dominant material removal mechanism shifts from extrusion to shear, and expansion of disordered atoms into the subsurface workpiece can be observed, creating nanocrystals on the machined surface.

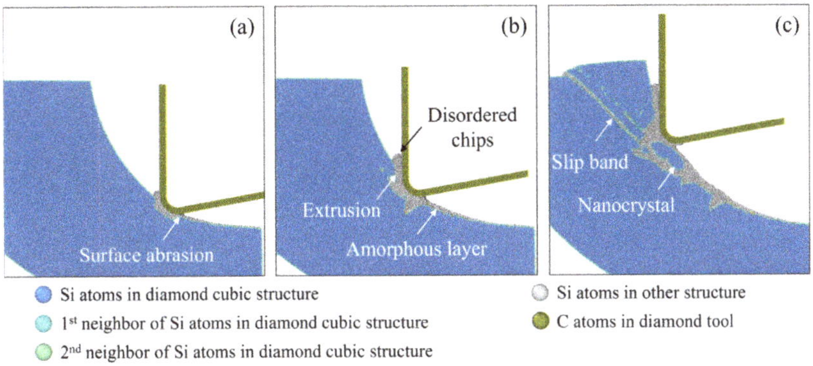

Figure 7. Snapshot of the crystal structure of workpiece atoms in different cutting stages from the simulation at 300 K with a tool edge radius of 5 nm. (**a**) 15 ps. (**b**) 25 ps. (**c**) 35 ps.

Figure 8 shows the crystal structure after machining under different conditions of tool edge radius and cutting temperature. During EVM of single-crystal Si, the dominant removal mechanism corresponds to the ratio between the tool edge radius and transient material removal thickness regarding the cutting direction. With increasing tool edge radius, the surface abrasion and extrusion stage is prolonged, forming more disordered atoms on the machined surface. And fewer nanocrystals are generated since slip motion and fracture of crystals are suppressed as the material removal thickness is decreased. As the temperature increases, the amorphous layer becomes less obvious, and more grains are left on the machined surface due to the suppressed high-pressure phase transition and enhanced dislocation-induced plastic deformation at high temperatures. Variation in the disordered atoms during machining under different conditions is shown in Figure 9. During the cutting process, the number of disordered atoms grows sharply from the initial contact stage as the material removal thickness increases. As the cutting tool advances, fewer disordered atoms are generated since the material removal mechanism switches into shear, and the transient material removal thickness decreases [24]. When machining at high temperatures (shown in Figure 9c,d), fewer disordered atoms are generated, and a slight decrease in disordered atoms can be observed in the shear stage, which can be attributed to the recrystallization process [37]. Therefore, the extension of disordered atoms in subsurface workpieces can be narrow and discontinuous under higher temperatures, as shown in Figure 9.

For materials such as single-crystal Si, recrystallization of disordered phases into crystal structures can be observed during high-temperature machining [38]. Figure 10 shows the recrystallization process when cutting at 1200 K with a tool edge radius of 25 nm, where the recrystallized atoms were identified by the CNA. It is observed that the recrystallization process is most obvious in the initial stage after machining, where the compressive stress remains in the deformed region since the transition of disordered Si into a cubic diamond structure can be promoted by applying compression and appropriate temperature [39]. Figure 11 shows the percentage of recrystallization under different cutting conditions, which is defined as the ratio between recrystallized atoms and the total disordered atoms. It is observed that recrystallization is promoted by raising the cutting temperature and tool edge radius. For blunt tools, a smaller material removal thickness indicates less generation of disordered atoms, while the enlarged contact area facilitates compression of the workpiece material, which is responsible for the promoted recrystallization percentage.

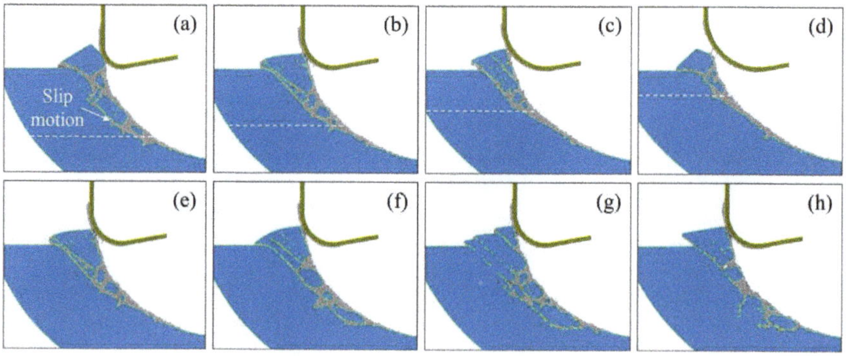

Figure 8. Crystal structure of workpiece atoms after cutting under different conditions. (**a–d**) cutting with tool edge radii of 10 nm, 15 nm, 20 nm, and 25 nm at 300 K, where the dashed line indicates the approximate position for the beginning of the slip motion. (**e–h**) cutting at 300 K, 600 K, 900 K, and 1200 K with a tool edge radius of 15 nm.

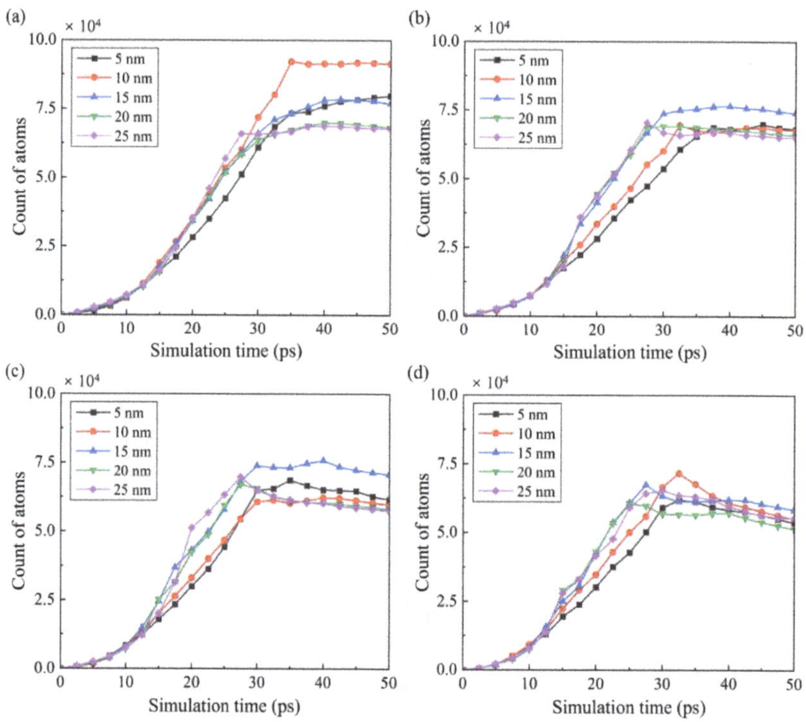

Figure 9. Variation in the disordered atoms cut at (**a**) 300 K, (**b**) 600 K, (**c**) 900 K, and (**d**) 1200 K by tools with different radii.

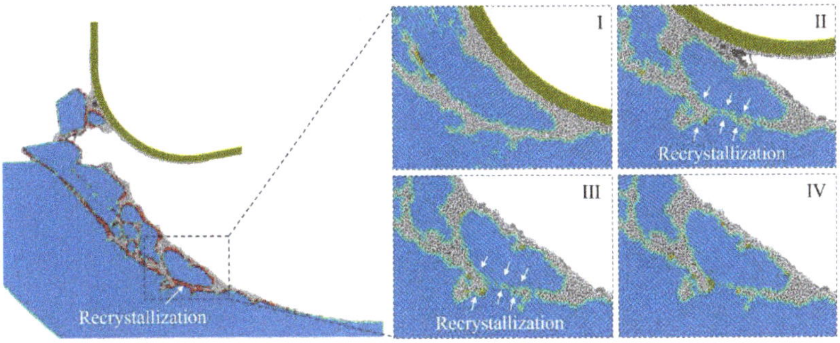

Figure 10. Illustration of the recrystallization during cutting at 1200 K with a tool edge radius of 25 nm, where the recrystallized atoms are in red color. (**I–IV**) shows the snapshots of the deformed region during different cutting stages.

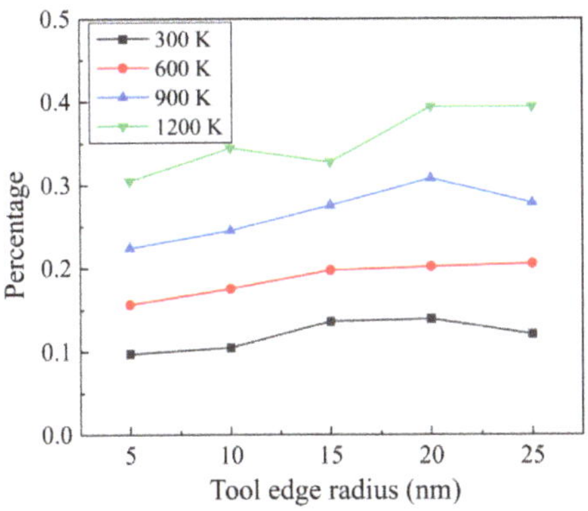

Figure 11. The percentage of recrystallization under different cutting conditions.

3.4. Stress Analysis

To further investigate the formation mechanism of subsurface damage, the internal stress in the workpiece was analyzed. The hydrostatic stress and von Mises stress in the workpiece were calculated based on the stress tensors from LAMMPS [40]:

$$\sigma_{hydrostatic} = \frac{1}{3}(\sigma_{xx} + \sigma_{yy} + \sigma_{zz}) \tag{5}$$

$$\sigma_{von\ Mises} = \sqrt{\frac{1}{2}\left(\left(\sigma_{xx} - \sigma_{yy}\right)^2 + \left(\sigma_{yy} - \sigma_{zz}\right)^2 + \left(\sigma_{zz} - \sigma_{xx}\right)^2 + 6\left(\tau_{xy}^2 + \tau_{yz}^2 + \tau_{zx}^2\right)\right)} \tag{6}$$

where σ_{xx}, σ_{yy}, σ_{zz}, τ_{xy}, τ_{xz}, and τ_{yz} are stress tensors. In this simulation, we used a spatial average of a cubic box of 2 nm to average the stress tensor, aiming to eliminate the fluctuation in the numerical calculation. Figure 12 shows the distribution of hydrostatic stress and von Mises stress in different cutting stages, where the workpiece was cut at 300 K by a tool with a 5 nm radius. Negative hydrostatic stress indicates compression, while positive hydrostatic stress suggests tension. In the initial contact stage, a high compressive zone is formed near the tool edge, causing surface abrasion and extrusion by introducing the high-pressure phase transition. A slight increase in tensile stress is observed behind the contact region due to the tearing of the workpiece by the cutting tool motion. As the material removal thickness increases, the compressive region is enlarged, and a shear region is formed due to the extrusion of disordered atoms. When the cutting process comes into the shear stage, variation in the cutting direction promotes the pulling-up motion, which contributes to the increase in tensile stress behind the contact region and in the subsurface workpiece. At the end of cutting, the internal stress decreases due to the detachment of crystal grains.

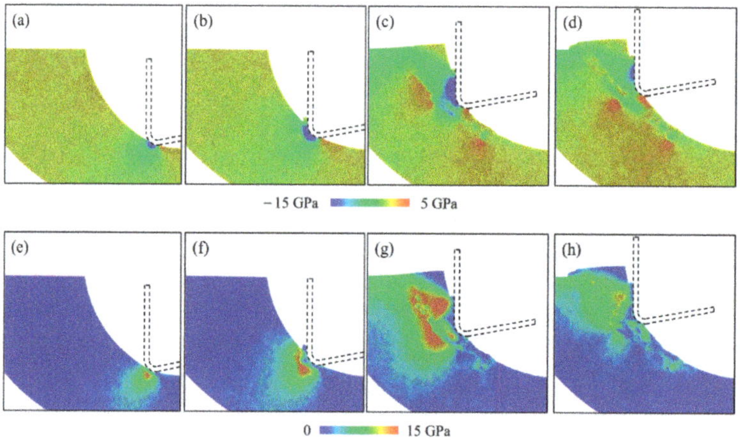

Figure 12. Distribution of the (**a**–**d**) hydrostatic stress and (**e**–**h**) von Mises stress in different cutting stages.

To investigate the variation in internal stress during the cutting stage, the average stress in a deformed region during the cutting stage under different conditions is calculated. The deformed region is defined as a box of 20 nm centered at T_c, as shown in Figure 13. The calculated average hydrostatic stress and von Mises stress are shown in Figures 14 and 15. During the cutting process, the hydrostatic stress increases as the material removal thickness increases from the initial contact stage. As the pulling-up motion increases and the detachment of crystal grains increases, the compressive stress decreases at the later cutting stage. Similar trends can be observed for the von Mises stress shown in Figure 15. As the cutting temperature increases, both hydrostatic stress and von Mises stress show an apparent decrease owing to the thermal softening of the workpiece material. For a workpiece machined by a blunt tool, an increase in internal stress is observed as the contact area is enlarged. In addition, when the tool edge radius increases, the internal stress reaches its maximum earlier, which is attributed to the fact that the detachment of the deformed region is promoted by a blunt tool. While this phenomenon becomes less apparent when the cutting temperature increases, since the enhanced plastic deformability at high temperature facilitates the slip motion and detachment of the crystal grains.

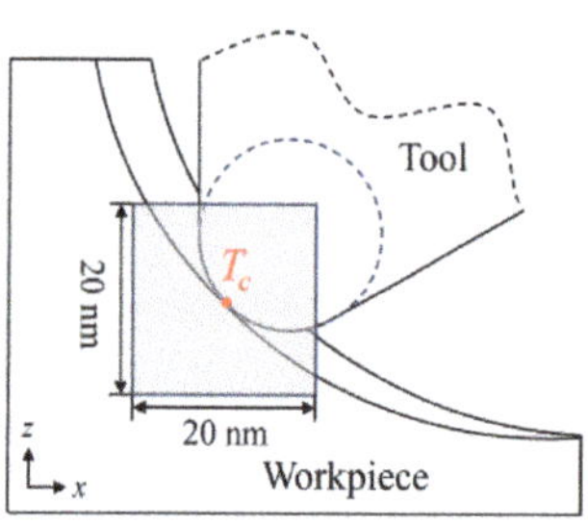

Figure 13. Definition of the deformed region for average stress calculation.

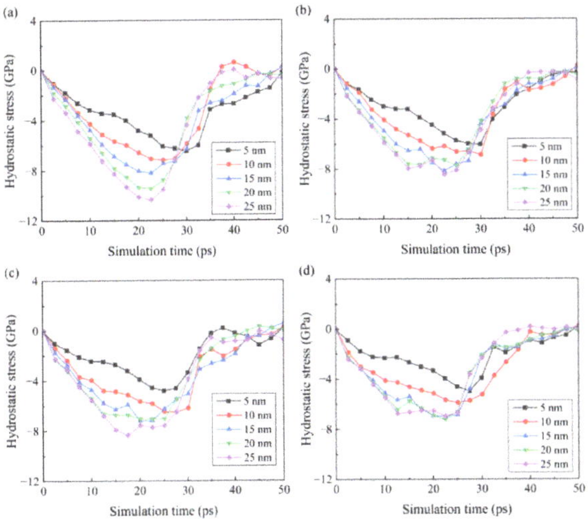

Figure 14. Variation in the hydrostatic stress and in the workpiece under different temperatures: (**a**) 300 K, (**b**) 600 K, (**c**) 900 K, and (**d**) 1200 K.

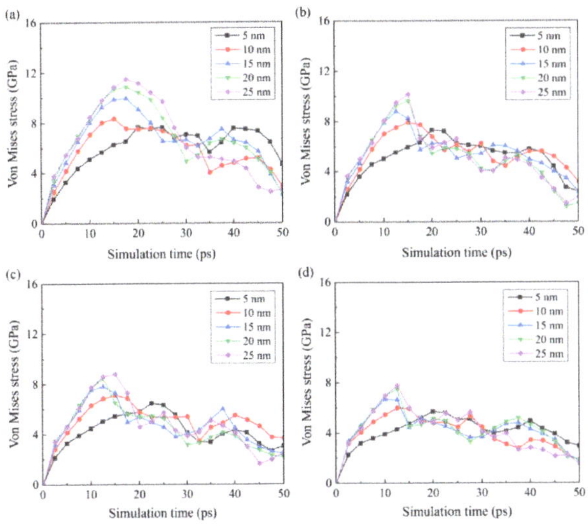

Figure 15. Variation in the von Mises stress and in the workpiece under different temperatures: (**a**) 300 K, (**b**) 600 K, (**c**) 900 K, and (**d**) 1200 K.

4. Conclusions

In this research, the material removal mechanism and subsurface damage generation in laser-vibration-assisted machining of single-crystal Si were examined using MD simulation. The impact of cutting temperature and tool edge radius on the machining mechanism was discussed. The following is a summary of the key findings:

(1) In one vibration cycle, surface swelling gradually becomes obvious after the initial contact due to the rise in material removal thickness, while it becomes less apparent at the end of the contact stage. When cutting at room temperature, the surface swelling is more obvious, and the surface roughness is increased when tools with a large edge radius are adopted. As the cutting temperature increases, a decrease in surface roughness is observed when the tool edge radius is increased.

(2) As the tool edge radius increases, more surface abrasion and extrusion removal can be observed during the cutting stage. As the cutting temperature increases, the plastic deformation in the subsurface workpiece is promoted. In the surface abrasion and extrusion stage, obvious plastic deformation patterns are observed. During the shear stage, more slip bands are observed with voids formed in the subsurface workpiece.

(3) With increasing tool edge radius, more disordered atoms are generated on the machined surface since the surface abrasion and extrusion stage is prolonged. As the temperature increases, the amorphous layer becomes inapparent, and more grains are left on the machined surface, which can be attributed to the enhanced dislocation-induced plastic deformation. Furthermore, recrystallization is promoted by raising the cutting temperature and tool edge radius.

(4) As the cutting temperature increases, both hydrostatic stress and von Mises stress show an apparent decrease owing to the thermal softening of the workpiece material. For workpieces machined by a blunt tool, an increase in internal stress is observed, and the internal stress reaches a maximum earlier. While this phenomenon becomes less apparent when the cutting temperature increases, since the enhanced plastic deformability facilitates the detachment of the crystal grains.

Author Contributions: Conceptualization, J.C. and Z.S.; methodology, J.C.; software, Y.Z.; validation, Y.Z. and J.K.; formal analysis, J.K.; investigation, J.C.; resources, C.C.; data curation, J.H. and A.X.; writing—original draft preparation, J.C.; writing—review and editing, J.C. and Y.Z.; visualization, Z.W. and Z.S.; supervision, Y.Y. and Z.S.; project administration, J.C.; funding acquisition, Z.W. and J.C. All authors have read and agreed to the published version of the manuscript.

Funding: This research was funded by the Doctoral Research Initiation Fund of Hubei University of Technology (No. XJ2024001101), the Hubei Key Laboratory of Modern Manufacturing Quality Engineering (KFJJ-2025009), and the National Natural Science Foundation of China under Grant number 52205522.

Data Availability Statement: The original contributions presented in the study are included in the article; further inquiries can be directed to the corresponding author.

Conflicts of Interest: The authors declare no conflicts of interest. Chen Chen, Jifei He, and Aijiang Xu are employees of Jiangsu Nuclear Power Corporation. The paper reflects the views of the scientists and not the company.

References

1. Kim, K.S.; Seo, S.; Kwon, J.; Lee, D.; Kim, C.; Ryu, J.E.; Kim, J.; Suh, J.M.; Jung, H.G.; Jo, Y.; et al. Growth-based monolithic 3D integration of single-crystal 2D semiconductors. *Nature* **2024**, *636*, 615–621. [CrossRef]
2. Priolo, F.; Gregorkiewicz, T.; Galli, M.; Krauss, T.F. Silicon nanostructures for photonics and photovoltaics. *Nat. Nanotechnol.* **2014**, *9*, 19–32. [CrossRef]
3. Du, H.; Zhao, D.; Liu, C.; Chen, H.; To, S. Laser-assisted slow tool servo diamond turning of single-crystal silicon for fabricating micro-lens arrays. *J. Manuf. Process.* **2024**, *127*, 174–183. [CrossRef]
4. Li, C.; Wang, K.; Zakharov, O.; Cui, H.; Wu, M.; Zhao, T.; Yan, Y.; Geng, Y. Damage evolution mechanism and low-damage grinding technology of silicon carbide ceramics. *Int. J. Extrem. Manuf.* **2025**, *7*, 022015. [CrossRef]
5. Wang, R.; Cheung, C.F.; Zang, Y.; Wang, C.; Liu, C. Material removal rate optimization with bayesian optimized differential evolution based on deep learning in robotic polishing. *J. Manuf. Syst.* **2025**, *78*, 178–186. [CrossRef]
6. Liu, C.; Chen, X.; Zhang, J.; Zhang, J.; Chu, J.; Xiao, J.; Xu, J. Molecular dynamic simulation of tool groove wear in nanoscale cutting of silicon. *AIP Adv.* **2020**, *10*, 015327. [CrossRef]
7. Chen, X.; Liu, C.; Ke, J.; Zhang, J.; Shu, X.; Xu, J. Subsurface damage and phase transformation in laser-assisted nanometric cutting of single crystal silicon. *Mater. Des.* **2020**, *190*, 108524. [CrossRef]
8. Ke, J.; Chen, X.; Liu, C.; Zhang, J.; Yang, H.; Xu, J. Enhancing the ductile machinability of single-crystal silicon by laser-assisted diamond cutting. *Int. J. Adv. Manuf. Technol.* **2022**, *118*, 3265–3282. [CrossRef]
9. Skelton, R.C. Turning with an oscillating tool. *Int. J. Mach. Tool Des. Res.* **1968**, *8*, 239–259. [CrossRef]

10. Shamoto, E.; Moriwaki, T. Study on Elliptical Vibration Cutting. *CIRP Ann.* **1994**, *43*, 35–38. [CrossRef]
11. Moriwaki, T.; Shamoto, E. Ultrasonic Elliptical Vibration Cutting. *CIRP Ann. Manuf. Technol.* **1995**, *44*, 31–34. [CrossRef]
12. Zhang, J.; Zhang, J.; Liu, C.; Chen, X.; Xiao, J.; Xu, J. Machinability of single crystal calcium fluoride by applying elliptical vibration diamond cutting. *Precis. Eng.* **2020**, *66*, 306–314. [CrossRef]
13. Yip, W.S.; To, S. An application of eddy current damping effect on single point diamond turning of titanium alloys. *J. Phys. D Appl. Phys.* **2017**, *50*, 435002. [CrossRef]
14. Hatefi, S.; Abou-El-Hossein, K. Review of magnetic-assisted single-point diamond turning for ultra-high-precision optical component manufacturing. *Int. J. Adv. Manuf. Technol.* **2022**, *120*, 1591–1607. [CrossRef]
15. Xing, Y.; Liu, Y.; Yin, T.; Li, D.; Sun, Z.; Xue, C.; Yip, W.S.; To, S. Magnetic and ultrasonic vibration dual-field assisted ultra-precision diamond cutting of high-entropy alloys. *Int. J. Mach. Tools Manuf.* **2024**, *202*, 104208. [CrossRef]
16. Ke, J.; Zhang, J.; Chen, X.; Liu, C.; Long, G.; Sun, H.; Xu, J. Investigation on the material removal mechanism in ion implantation-assisted elliptical vibration cutting of hard and brittle material. *Int. J. Mach. Tools Manuf.* **2024**, *203*, 104220. [CrossRef]
17. Zhang, J.; Li, X.; Ma, S.; Lu, Y.; Tian, H.; Xiao, J.; Xu, J. High efficiency fabrication of Si microlenses by applying in-situ laser and ultrasonic vibration hybrid diamond cutting. *Precis. Eng.* **2025**, *93*, 324–333. [CrossRef]
18. Li, C.; Hu, Y.; Zhang, F.; Geng, Y.; Meng, B. Molecular dynamics simulation of laser assisted grinding of GaN crystals. *Int. J. Mech. Sci.* **2023**, *239*, 107856. [CrossRef]
19. Liu, C.; To, S.; Sheng, X.; Xu, J. Molecular dynamics simulation on crystal defects of single-crystal silicon during elliptical vibration cutting. *Int. J. Mech. Sci.* **2023**, *244*, 108072. [CrossRef]
20. Wang, J.; Zhang, X.; Fang, F. Molecular dynamics study on nanometric cutting of ion implanted silicon. *Comput. Mater. Sci.* **2016**, *117*, 240–250. [CrossRef]
21. Dai, H.; Li, S.; Chen, G. Comparison of subsurface damages on mono-crystalline silicon between traditional nanoscale machining and laser-assisted nanoscale machining via molecular dynamics simulation. *Nucl. Instrum. Methods Phys. Res. Sect. B Beam Interact. Mater. At.* **2018**, *414*, 61–67. [CrossRef]
22. Liu, C.; Ke, J.; Yin, T.; Yip, W.S.; Zhang, J.; To, S.; Xu, J. Cutting mechanism of reaction-bonded silicon carbide in laser-assisted ultra-precision machining. *Int. J. Mach. Tools Manuf.* **2024**, *203*, 104219. [CrossRef]
23. Liu, C.; Zhang, J.; Zhang, J.; Chen, X.; Xiao, J.; Xu, J. A simulation investigation on elliptical vibration cutting of single-crystal silicon. *Int. J. Adv. Manuf. Technol.* **2020**, *108*, 2231–2243. [CrossRef]
24. Liu, C.; Zhang, J.; Zhang, J.; Chu, J.; Chen, X.; Xiao, J.; Xu, J. Numerical investigation on material removal mechanism in elliptical vibration cutting of single-crystal silicon. *Mater. Sci. Semicond. Process.* **2021**, *134*, 106019. [CrossRef]
25. Liu, C.; He, W.; Chu, J.; Zhang, J.; Chen, X.; Xiao, J.; Xu, J. Molecular Dynamics Simulation on Cutting Mechanism in the Hybrid Machining Process of Single-Crystal Silicon. *Nanoscale Res. Lett.* **2021**, *16*, 66. [CrossRef]
26. Plimpton, S. Fast parallel algorithms for short-range molecular dynamics. *J. Comput. Phys.* **1995**, *117*, 1–19. [CrossRef]
27. Stukowski, A. Visualization and analysis of atomistic simulation data with OVITO–the Open Visualization Tool. *Model. Simul. Mater. Sci. Eng.* **2010**, *18*, 015012. [CrossRef]
28. Liu, C.; Chu, J.; Chen, X.; Xiao, J.; Xu, J. Molecular dynamics simulation on structure evolution of silica glass in nano-cutting at high temperature. *Mol. Simul.* **2020**, *46*, 957–965. [CrossRef]
29. Liu, C.; Yip, W.S.; To, S.; Chen, B.; Xu, J. Numerical Investigation on the Effects of Grain Size and Grinding Depth on Nano-Grinding of Cadmium Telluride Using Molecular Dynamics Simulation. *Nanomaterials* **2023**, *13*, 2670. [CrossRef] [PubMed]
30. Erhart, P.; Albe, K. Analytical potential for atomistic simulations of silicon, carbon, and silicon carbide. *Phys. Rev. B* **2005**, *71*, 035211. [CrossRef]
31. Guo, X.; Li, Q.; Liu, T.; Zhai, C.; Kang, R.; Jin, Z. Molecular dynamics study on the thickness of damage layer in multiple grinding of monocrystalline silicon. *Mater. Sci. Semicond. Process.* **2016**, *51*, 15–19. [CrossRef]
32. Dai, H.; Chen, G.; Zhou, C.; Fang, Q.; Fei, X. A numerical study of ultraprecision machining of monocrystalline silicon with laser nano-structured diamond tools by atomistic simulation. *Appl. Surf. Sci.* **2017**, *393*, 405–416. [CrossRef]
33. Cai, M.B.; Li, X.P.; Rahman, M. Characteristics of "dynamic hard particles" in nanoscale ductile mode cutting of monocrystalline silicon with diamond tools in relation to tool groove wear. *Wear* **2007**, *263*, 1459–1466. [CrossRef]
34. Chen, X.; Chu, J.; Zhu, Z.; Liu, C. Effect of temperature on surface morphology of single-crystal silicon in nanometric cutting. *Appl. Surf. Sci.* **2025**, *684*, 161957. [CrossRef]
35. Liu, C.; Chu, J.; Zhang, J.; Zhang, J.; Chen, X.; Xiao, J.; Xu, J. Effect of tool rake angle on the material removal mechanism transition of single-crystal silicon: A molecular dynamics study. *Int. J. Adv. Manuf. Technol.* **2021**, *115*, 3631–3644. [CrossRef]
36. Maras, E.; Trushin, O.; Stukowski, A.; Ala-Nissila, T.; Jónsson, H. Global transition path search for dislocation formation in Ge on Si(001). *Comput. Phys. Commun.* **2016**, *205*, 13–21. [CrossRef]
37. Liu, C.; Yip, W.S.; Chen, J.; Wang, R.; Xu, J.; To, S. Atomic simulation of the temperature effect on fabrication mechanism of micro-structured surface on single-crystal silicon. *J. Manuf. Process.* **2025**, *133*, 238–248. [CrossRef]

38. Liu, C.; Xu, W.; Zhang, J.; Xiao, J.; Chen, X.; Xu, J. Numerical investigation on the temperature effect in nanometric cutting o polycrystalline silicon. *Int. J. Mech. Sci.* **2022**, *220*, 107172. [CrossRef]
39. Liu, C.; Chen, X.; Ke, J.; She, Z.; Zhang, J.; Xiao, J.; Xu, J. Numerical investigation on subsurface damage in nanometric cutting o single-crystal silicon at elevated temperatures. *J. Manuf. Process.* **2021**, *68*, 1060–1071. [CrossRef]
40. Goel, S.; Luo, X.; Agrawal, A.; Reuben, R.L. Diamond machining of silicon: A review of advances in molecular dynamic simulation. *Int. J. Mach. Tools Manuf.* **2015**, *88*, 131–164. [CrossRef]

micromachines

MDPI

rticle

'OLO-LA: Prototype-Based Vision–Language Alignment for ;ilicon Wafer Defect Pattern Detection

iyue Wang [1], Yichen Yang [2], Jianning Chu [3], Yikai Zang [4], Zhongdi She [4], Weikang Fang [1,*] and Ruoxin Wang [5,*]

1 School of Intelligent Manufacturing, Jianghan University, Wuhan 430056, China; wangziyue0829@126.com
2 Marine Design and Research Institute of China, Shanghai 200011, China
3 Hubei Key Laboratory of Modern Manufacturing Quality Engineering, School of Mechanical Engineering, Hubei University of Technology, Wuhan 430068, China
4 State Key Laboratory of Intelligent Manufacturing Equipment and Technology, School of Mechanical Science and Engineering, Huazhong University of Science and Technology, Wuhan 430074, China
5 Department of Mechanical Engineering, University of Maryland College Park, College Park, MD 20742, USA
* Correspondence: hustfwk@163.com (W.F.); rwang135@umd.edu (R.W.)

Abstract

With the rapid development of semiconductor manufacturing technology, methods to effectively control the production process, reduce variation in the manufacturing process, and improve the yield rate represent important competitive factors for wafer factories. Wafer bin maps, a method for characterizing wafer defect patterns, provide valuable information for engineers to quickly identify potential root causes through accurate pattern recognition. Vision-based deep learning approaches rely on visual patterns to achieve robust performance. However, they rarely exploit the rich semantic information embedded in defect descriptions, limiting interpretability and generalization. To address this gap, we propose YOLO-LA, a lightweight prototype-based vision–language alignment framework that integrates a pretrained frozen YOLO backbone with a frozen text encoder to enhance wafer defect recognition. A learnable projection head is introduced to map visual features into a shared embedding space, enabling classification through cosine similarity Experimental results on the WM-811K dataset demonstrate that YOLO-LA consistently improves classification accuracy across different backbones while introducing minimal additional parameters. In particular, YOLOv12 achieves the fastest speed while maintaining competitive accuracy, whereas YOLOv10 benefits most from semantic prototype alignment. The proposed framework is lightweight and suitable for real-time industrial wafer inspection systems.

Keywords: semiconductor; defect detection; wafer bin map; YOLO; vision language model

1. Introduction

Semiconductor components require high manufacturing quality with minimal surface defects [1,2]. To overcome the brittleness and hardness of semiconductor materials, complex manufacturing processes have been proposed [3–5], and the intricacy of manufacturing defects in wafers produced during these processes has increased. A wafer bin map (WBM) [6] visually represents the test outcomes for each chip on a wafer, based on the chip's probe test failure mode and its position (die). The chip probe test, an essential final assessment after the entire manufacturing sequence, evaluates the performance and functionality of each chip. During this test, each die is subjected to multiple probe test modes, with the first failure mode being recorded as the bin result. Throughout the wafer fabrication process, various manufacturing issues can lead to multiple defective dies on a wafer. These defects often cluster in specific areas on the wafer, forming spatial patterns known as gross failure areas (GFAs). These patterns, which include common types such as ring, scratch, loc, and center, are indicative of process-related issues and provide valuable data for improving yield and quality [7]. Classifying GFAs is crucial for engineers to identify and address problems in the production process, thereby reducing costs and enhancing yield. As production environments become more intricate, the need for automated WBM GFA classification has become increasingly important.

Deep learning-based object detection outperforms traditional algorithms and classification networks in terms of generalization ability and localization precision, leading to superior overall performance. Object detection algorithms can be categorized into two types: single-stage networks, exemplified by YOLO [8], SSD [9], and RetinaNet [10], and two-stage networks, represented by Faster R-CNN [11]. While two-stage networks prioritize accuracy over speed by separating localization and recognition, single-stage networks simultaneously perform both tasks, resulting in faster detection speeds. Although these vision-based models achieve competitive performance, they only rely on pixel-level information and ignore rich semantic knowledge inherent in defect pattern descriptions.

Meanwhile, recent progress in vision–language models (VLMs), such as CLIP-style architectures [12], demonstrates that aligning visual representations with natural-language semantics significantly improves generalization and interpretability. Much VLM research has been successfully applied in industrial defect detection. AnomalyGPT was proposed by Gu et al. [13] to address industrial anomaly detection by integrating fine-grained visual decoding and multi-turn dialog capability. It achieved state-of-the-art image- and pixel-level performance on MVTec-AD with strong few-shot generalization from only one normal example. Qian et al. [14] proposed a contrastive cross-modal training framework named CLAD that adopted large vision–language models to align visual and textual representations for industrial anomaly detection and localization. By jointly improving image-level detection and pixel-level localization on benchmarks such as MVTec-AD and VisA, it demonstrates that cross-modal contrastive learning can enhance both performance and interpretability in industrial inspection tasks. Cao et al. [15] developed a model named AnomalyVLM to address zero-shot industrial anomaly detection In AnomalyVLM, product standards were regarded as a substitute for reference images, enabling vision–language models to reason about normality and abnormality without category-specific training data.

However, integrating VLM concepts into industrial wafer map analysis remains largely unexplored. In particular, wafer defect classes naturally correspond to concise textual descriptions, suggesting an opportunity to leverage language as a source of domain knowledge. Therefore, in this paper, we introduce YOLO-LA, a prototype-based vision–language alignment framework, in which a trained YOLO backbone serves as a visual encoder, while a frozen text encoder generates semantic prototypes from defect descriptions. A lightweight learnable projection head aligns the YOLO visual embedding with the text embedding space, enabling wafer classification via cosine similarity to the textual prototypes.

2. YOLO Architectures and Baseline Model

YOLO, which stands for "You Only Look Once," is a state-of-the-art, real-time object detection system that has revolutionized the field of computer vision. Developed by Joseph Redmon and his collaborators [8], YOLO is designed to detect and classify objects within images or video streams with remarkable speed and accuracy. Unlike traditional object detection methods that apply a classifier to various parts of an image, YOLO reframes object detection as a single regression problem, directly predicting bounding boxes and

https://doi.org/10.3390/mi1701006

class probabilities from full images in one evaluation. This approach allows YOLO to process images in real time, making it highly efficient for applications that require fast and accurate object detection, such as autonomous driving, surveillance, and augmented reality. YOLO's architecture is based on a convolutional neural network (CNN) that divides the input image into a grid and predicts bounding boxes and probabilities for each grid cell, enabling it to detect multiple objects of different classes simultaneously. Over the years, YOLO has undergone several iterations, each improving upon the previous in terms of speed, accuracy, and robustness, solidifying its position as a leading framework in the object detection domain.

2.1. YOLOv8

YOLOv8 was proposed by Jocher et al. [16]. In YOLOv8, the backbone is responsible for extracting hierarchical features from the input image. It consists of several sequential layers, including CBS, C2f, and SPPE. CBS typically stands for Convolution + BatchNorm + SiLU activation. It is used to process feature maps efficiently. C2f is a modified bottleneck structure, designed for feature extraction with fewer parameters. SPPF (Spatial Pyramid Pooling—Fast) enhances the receptive field and captures multi-scale features at the end of the backbone. The neck is designed to aggregate and fuse features from different stages of the backbone, helping detect objects at various scales. It includes C2f blocks for feature processing, Upsample operations to increase spatial resolution, and Concat operations to fuse feature maps from different layers. This part resembles a Feature Pyramid Network (FPN) combined with a Path Aggregation Network (PAN), enabling both top–down and bottom–up feature fusion. The head is responsible for final predictions.

2.2. YOLOv10

YOLOv10 [17] is the most representative model in YOLO series models. YOLOv10's architecture leverages the strengths of its predecessors while incorporating several groundbreaking innovations. In contrast, in YOLOv10, the backbone utilizes an advanced version of C2F (cross-stage partial with two-fusion layers), called C2fCIB, which optimizes gradient flow and minimizes computational redundancy [18]. The neck module is designed to aggregate features from various scales, which employs PANet (Path Aggregation Network) layers to facilitate effective multi-scale feature fusion and pass them to the head.

2.3. YOLOv11

YOLOv11 [19] builds upon the foundation of YOLOv10, retaining its core design principles such as the modular backbone–neck–head architecture and the use of efficient components like CBS and SPPF blocks. Both versions adopt a one-to-many prediction strategy in the head, enabling each spatial location to generate multiple object detections. However, YOLOv11 introduces several key improvements over YOLOv10. YOLOv11 introduces a novel C3K2 (cross-stage partial with kernel size 2) module, which is an improved version of C2f, which combines the CSP concept with a smaller convolution kernel. By splitting the feature map into two parts, one is passed directly, while the other undergoes multiple bottleneck convolutions before being fused with the first part. This preserves rich features while reducing computation and parameter count, so as to improve computational efficiency and reduce inference latency. In addition, they also modified the PSA to C2PSA. In the C2PSA module, a convolutional layer with a kernel size of 1 is performed to split the feature map into two parts. One part enters the PSA module to highlight the response of key spatial regions, while the other part retains the original features. In the end, the two parts are concatenated and fused through another convolutional layer with a kernel size of 1.

https://doi.org/10.3390/mi17010067

2.4. YOLOv12

The YOLOv12 [20] architecture is composed of four key components, including a backbone module, neck module, and head module, where two techniques are applied to reduce computational cost. First, a residual efficient layer aggregation network (R-ELAN) in the backbone module was proposed to integrate features from different scales and reduce the computation cost and memory usage simultaneously. In addition, area attention was used in the neck module to reduce computation by narrowing the attention from global to local. The head part is the simple classification network, which uses a convolutional layer followed by an adaptive average pool layer. The last linear layer represents the output number of defect types.

3. YOLO-LA: Prototype-Based Vision–Language Alignment Framework

3.1. Overall Framework

A lightweight prototype-based vision–language alignment framework, termed YOLO-LA, is proposed, that enriches YOLO visual embeddings with semantic information extracted from natural-language defect descriptions [21]. Unlike standard CLIP pipelines that rely on large-scale image–text pretraining and open-vocabulary recognition, YOLO-LA operates on compact, topology-driven wafer maps and leverages frozen industrial visual backbones together with fixed semantic defect prototypes. YOLO-LA contains three major components, as shown in Figure 1: (1) a pretrained frozen YOLO backbone that produces visual embeddings of wafer maps, (2) a frozen text encoder that generates semantic prototypes from defect class descriptions, (3) a learnable projection head that aligns the visual and textual embedding spaces for prototype-based classification.

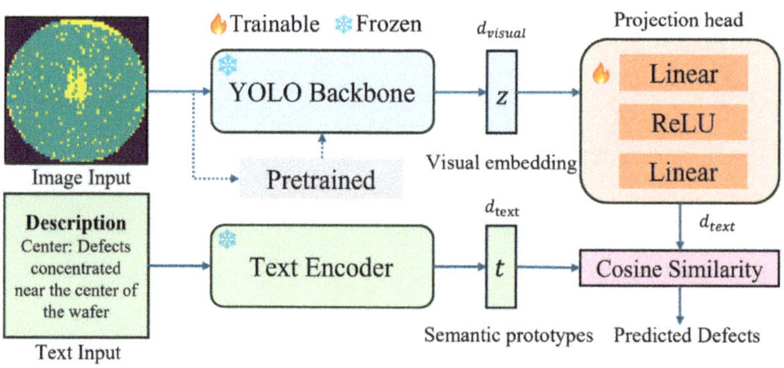

Figure 1. YOLO-LA framework. Given an input wafer map image, a pretrained YOLO backbone extracts visual features, which are mapped into the semantic embedding space via a lightweight projection head. Textual class descriptions are encoded into fixed semantic prototypes. Final classification is obtained by cosine similarity between visual embeddings and semantic prototypes. During training, the YOLO backbone is frozen unless otherwise specified.

Visual Embedding Extraction: After training a YOLO classifier on wafer map dataset, the YOLO backbone is frozen and regarded as a visual feature extractor. Given an input image $x \in R^{H \times W \times 3}$, the YOLO backbone produces a high-level feature map $F \in R^{M \times S \times S}$ as Equation (1). By applying a global average pooling (GAP) layer, a compact visual descriptor $z \in R^{d_{visual}}$ is obtained for alignment.

$$F = GAP\left(f_{yolo}(x)\right) \tag{1}$$

$$z = GAP(F) \tag{2}$$

https://doi.org/10.3390/mi17010067

Here, H and W are the height and width of the input image, M and S are the number of channels and feature map size of the last feature layer of YOLO backbone, and d_{visual} is the dimension of visual embedding.

Text Prototype Construction from Defect Descriptions: Each wafer defect class naturally corresponds to a short textual description. The defects and their descriptions are listed in Table 1. Each description is encoded to a text embedding $t_c \in R^{d_{text}}$ by a frozen sentence-level language model f_{text} as Equation (3):

$$t_c = f_{text}(d_c),\ c = \{1, 2, \dots C\} \tag{3}$$

where d_{text} is the dimension of textual embedding, C is the number of defect types, and d_c is a description of c-th defect type. Then, ℓ_2 normalization is applied to obtain scaled text embeddings $\tilde{t}_c$. The final semantic prototype $t \in R^{C \times d_{text}}$ is generated by stacking all text embeddings as Equation (4) to serve as semantic anchors.

$$t = [\tilde{t}_1; \tilde{t}_2, \dots, \tilde{t}_C] \tag{4}$$

Table 1. Textual description of defects.

Defect Type	Description
Center	Defects concentrated near the center of the wafer
Donut	Donut-shaped defect pattern on the wafer
Edge_Ring	Ring-shaped defects along the wafer edge
Edge_Loc	Localized defects near the wafer edge
Loc	Localized defect cluster in a small region
Random	Randomly scattered defects on the wafer
Scratch	Linear scratch-like defects across the wafer
Near_full	Defects covering most of the wafer area

Projection Head for Visual–Language Alignment: To align YOLO visual embedding z with language-based defect descriptions, a lightweight trainable two-layer MLP projection head is designed, as shown in Equations (5) and (6).

$$h = Linear(ReLU(Linear(z))) \in R^{d_h} \tag{5}$$

$$v = Linear(h) \in R^{d_{text}} \tag{6}$$

Here, $Linear$ and $ReLU$ are the linear projection layer and activation function; d_h is the dimension of the hidden layer; h and v are hidden embedding and aligned visual embedding. In this way, only the parameters of the projection head (<0.2 M parameters) are trained, while the YOLO backbone and text encoder remain frozen, ensuring high efficiency.

In the end, the final prediction is calculated by cosine similarity as Equation (7),

$$\hat{y} = \underset{c}{\text{argmax}} \tilde{v}^T \tilde{t}_c \tag{7}$$

where $\tilde{v}$ is the ℓ_2-normalized v.

https://doi.org/10.3390/mi17010067

3.2. Loss Function and Metrics

During training, a cross-entropy loss is utilized to optimize the projection head parameters as shown in Equation (8):

$$\mathcal{L} = -\sum_{i} log \frac{\exp(\tilde{v}^T \tilde{t}_c)}{\sum_c \exp(\tilde{v}^T \tilde{t}_c)} \tag{8}$$

In the paper, Accuracy [22] is adopted to evaluate the classification accuracy performance. Accuracy is defined as Equation (9):

$$Accuracy = \frac{TP + TN}{TP + TN + FP + FN} \tag{9}$$

where TP is true positive, FN is false negative, TN is true negative, and FP is false positive. Top-N accuracy refers to the proportion of instances where the correct label is among the N predictions with the highest probabilities.

4. Experiments and Results

4.1. Dataset and Experimental Settings

For training purposes, we utilized the WM-811K dataset [23], a comprehensive collection of 811,457 wafers sourced from real-world fabrication environments. The WM-811K dataset comprises 172,950 labeled samples and 639,507 unlabeled samples, with image sizes varying from 6 × 21 to 300 × 202 pixels. The labeled images are categorized into nine distinct types, including eight defect pattern types and a 'None' type representing normal wafers. In this study, we converted 25,519 labeled images with defect patterns into 3-channel images with a uniform size of 128 × 128 pixels. The dataset is first divided into training and test sets using an 80:20 stratified split. The training set is further split into training and validation subsets for model selection, while the test set is strictly held out for final evaluation. All splits are generated using a fixed random seed (42) and shared across all experiments. The eight defect types, including Center, Donut, Edge_Loc, Edge_Ring, Loc, Random, Scratch, and Near_full, are shown in Figure 2. In the pretraining process of the YOLO backbone, we set the training epoch as 50 and the optimizer as stochastic gradient descent with a learning rate of 0.01 and momentum of 0.9. The batch size is set as 16. All YOLO backbones used in this work are initialized with ImageNet-pretrained weights, following standard practice for classification tasks. The backbone networks are first trained using the native YOLO classification objective, which employs a SoftMax-based cross-entropy loss for multi-class classification. In the proposed YOLO-LA framework, the pretrained YOLO backbone is primarily used as a visual feature extractor. Unless otherwise specified in the ablation study, the backbone parameters are kept frozen, and only the lightweight projection head for semantic prototype alignment is optimized. This design ensures that the proposed method does not alter the original YOLO training objective but instead augments the learned visual representations through vision–language alignment. In YOLO-LA, MiniLM-L6-v2 is used for encoding each description, the training epoch is set to 100, the learning rate is 3×10^{-4}, and the batch size and image size remain the same as in YOLO backbone pretraining. The d_h in the projection head is 512. All experiments are run on a computer with an i7-9750H 2.60GHz CPU and NVIDIA GEFORCE RTX 2060 GPU.

https://doi.org/10.3390/mi1701006

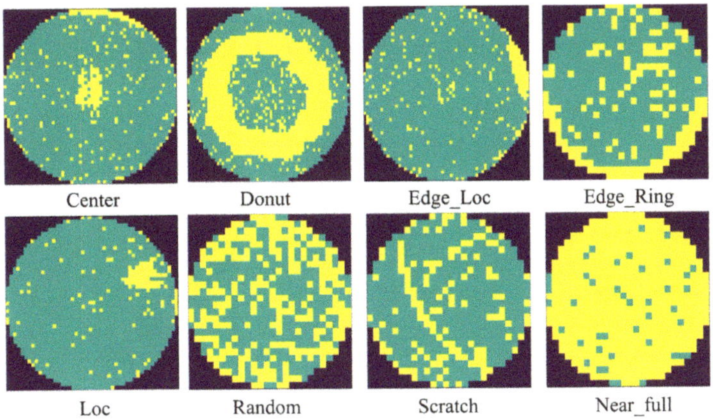

Figure 2. The eight types of defect pattern.

4.2. Results

First, we explore the YOLO backbone performance. Figure 3 shows confusion matrices. Figure 3a shows the result of YOLOv8, where the accuracy of some classes, such as Edge-Loc and Edge-Ring, was low, and there is significant misclassification of Random and Background. Many classes were less than 0.8, indicating relatively weak overall performance. Figure 3b demonstrates that YOLOv10 achieved accuracy between 0.85 and 0.99 in most defect types, significantly reducing confusion. In addition, background classification accuracy was near-perfect, slightly improving the overall accuracy. The results of YOLOv11 are shown in Figure 3c. The accuracy is high for most defect types, especially Near-full, Random, and Scratch, even approaching 1.0 for some types. However, defects with similar morphology, such as Edge_Loc and Edge_Ring, still exhibit confusion, although to a lesser extent than in YOLOv8 and YOLOv10. The results of YOLOv12 are shown in Figure 3d. The model achieves the best results on Center defect patterns, reaching 0.96, and the worst results on Loc defect patterns. The Loc defect pattern can be easily confused with Edge-Loc, and the Scratch defect pattern can be easily confused with Loc. The reason is that the YOLOv12 architectural design is consistent with more stable predictions on defect categories characterized by strong global spatial structures, such as Center and Edge_Ring, as reflected in the confusion matrix.

Then, we compared the YOLOv8-N, YOLOv10-N, YOLOv11-N, and YOLOv12-N architecture performance with the corresponding YOLO-LA model. As shown in Table 2, YOLO-LA can improve the YOLO performance whichever YOLO backbone is adopted. The YOLO-LA model based on the YOLOv10-N backbone network achieved the best performance, with the most significant improvement in semantic prototypes in YOLOv10-N, reaching 2.25%. YOLOv12-N achieved the fastest inference time (0.3 ms) and showed a competitive performance, also gaining +1.58%. The reason is that YOLOv12 includes enhanced operator fusion, reduced redundant computation, and more efficient feature aggregation, which collectively reduce inference latency. YOLOv11-N shows the smallest improvement (+0.39%). One reason is that the baseline YOLOv11 model already exhibits relatively balanced per-class performance, resulting in limited inter-class confusion and therefore less room for improvement through semantic alignment. These results indicate that semantic alignment should be viewed as a selective enhancement mechanism rather than a uniform performance booster, providing the greatest benefit for backbones and scenarios with higher baseline inter-class confusion. Overall, the proposed YOLO-

https://doi.org/10.3390/mi17010067

LA framework is backbone-agnostic and provides consistent performance enhancement without modifying or fine-tuning the YOLO classifier itself.

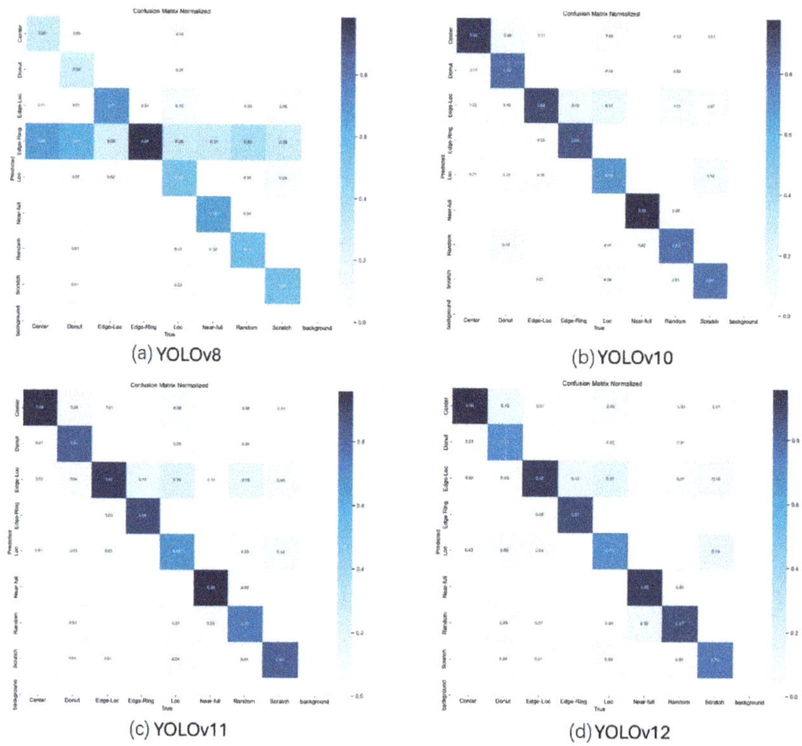

Figure 3. The confusion matrix of YOLOs.

Table 2. The performance comparison of different YOLOs. Bold numbers indicate the best performance for each column, and underlined numbers indicate the second-best performance.

Backbone	Params (M)	Inference Time (ms)	Accuracy % (YOLO)	Accuracy % (YOLO-LA)
YOLOv8-N	1.4	3.0	83.28	84.85 (+1.54)
YOLOv10-N	0.9	6.1	83.50	**85.75 (+2.25)**
YOLOv11-N	1.5	1.3	**83.61**	84.00 (+0.39)
YOLOv12-N	1.7	**0.3**	83.48	85.06 (+1.58)

In addition, Figure 4 presents representative classification examples comparing the baseline YOLO classifier and the proposed YOLO-LA model. Here, we used YOLOv10-N as the backbone. These cases show that YOLO-LA has a high ability to detect visually ambiguous defect patterns with high inter-class similarity. Specifically, the misclassified samples are all the extremely small effective resolution of wafer map images (approximately 26×26 pixels), which means purely visual models are prone to misinterpreting global structural patterns. YOLO-LA can introduce high-level structural priors from defect descriptions, leading to more robust classification under low-resolution conditions.

https://doi.org/10.3390/mi17010067

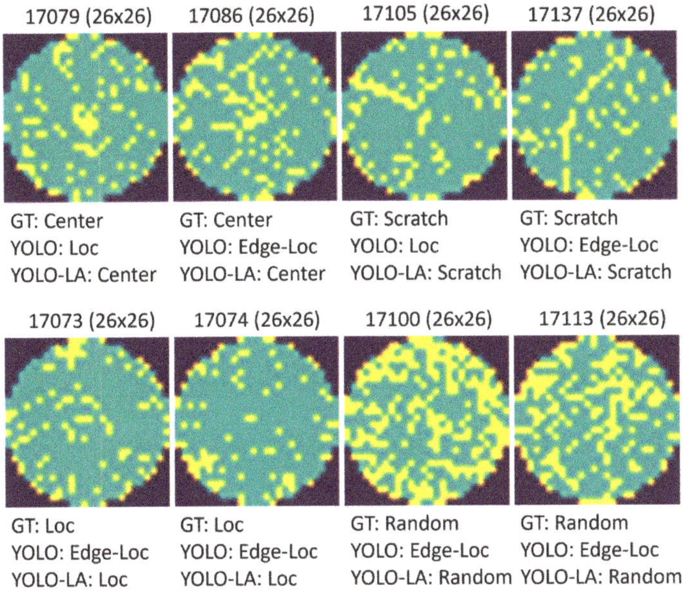

Figure 4. Visualization comparison results between YOLO and YOLO-LA.

4.3. Ablation Study

Text Encoder. To further explore the performance of YOLO-LA, we conducted a series of text encoder studies using YOLOv10 backbones (Table 3), such as MiniLM-L3 [24], MiniLM-L6 [24], MiniLM-L12 [24], Mpnet-Base [25], Roberta-Large [26], and E5-Base [27]. The MiniLM family provides a favorable trade-off between efficiency and semantic representation quality. Among them, MiniLM-L12 showed the best classification performance. MiniLM-L6 with 33 M parameters achieved a balanced configuration and was therefore adopted as the default text encoder in most experiments. The text encoder serves as a semantic anchor for high-level attributes (e.g., concentration, locality) rather than a linguistic reasoning engine. In contrast, E5-base employed a larger backbone with higher-dimensional embeddings and achieved the second-best performance. However, the increased embedding dimension cannot ensure better performance and slightly increases training costs. This indicates that the effectiveness of YOLO-LA does not rely on the expressive power of large language models but instead on introducing structured semantic priors into the visual decision process.

Table 3. The ablation study for various text encoders. Bold numbers indicate the best performance for each column, and underlined numbers indicate the second-best performance.

Text Encoder	d_{text}	Params (M)	Encoding Time (ms)	Accuracy (%)
MiniLM-L3	384	22	**18.72**	85.72
MiniLM-L6	384	33	<u>31.88</u>	85.75
MiniLM-L12	384	66	60.28	**85.90**
Mpnet-Base	768	110	59.79	85.67
RoBERTa-Large	1024	355	83.58	85.77
E5-base	768	110	46.68	<u>85.88</u>

Prompt Type: As shown in Table 4, we compared different prompt types, including base, detailed, and concise descriptions of defects. The detailed and concise descriptions for each defect are listed in Appendix A Tables A1 and A2. It is observed that the base prompt

https://doi.org/10.3390/mi17010067

achieved the best overall performance, with an accuracy of 84.44% and a macro-F1 of 51.16%, while also requiring the shortest training time. Macro-F1 is adopted to evaluate per-class performance by equally weighting the F1 score of each defect category. This suggests that semantic prototype alignment benefits mainly from high-level structural semantics rather than prompt verbosity and that both overly detailed and overly compressed descriptions can introduce redundant or ambiguous cues. Based on the overall accuracy, it can be concluded that this process is not dominated by specific keywords within the embeddings.

Table 4. The ablation studies for various prompt types.

Prompt Type	Training Time (s)	Accuracy (%)	Macro-F1 (%)
Base	327.30	84.44	51.16
Detailed	405.11	83.26	50.37
Concise	405.97	82.11	49.87

Training Strategy: We further explored the effect of different training strategies, including freezing the YOLO backbone, partially fine-tuning the last two blocks, and fully fine-tuning the entire network. As shown in Table 5, freezing the backbone consistently yields the best performance, achieving 85.31% accuracy and 51.88% macro-F1, while also being the most computationally efficient. In contrast, partial and full fine-tuning lead to longer training times and noticeable performance degradation. This indicates that, under low-resolution wafer map settings, aggressive fine-tuning may disturb pretrained visual representations and increase overfitting risk.

Table 5. The ablation studies for different training strategies.

Training Strategy	Training Time (s)	Accuracy (%)	Macro-F1 (%)
Freeze	418.42	85.31	51.88
Last-2	464.03	83.31	50.38
All	564.45	79.42	47.96

Projection Head: The projection head architecture was further ablated in terms of activation function, hidden dimension, and network depth. The results in Tables 6 and 7 show that increasing the hidden dimension from 256 to 512 consistently improves both accuracy and macro-F1, while larger depths bring marginal or unstable gains. Among different activation functions, GELU slightly outperforms ReLU for smaller hidden dimensions, whereas their performance becomes comparable at higher dimensions. Overall, a two-layer projection head with 512 hidden units provides the best balance between performance and training efficiency, supporting the use of a lightweight alignment module.

Table 6. The ablation study for different activation functions and hidden dimensions.

Activation Function	Dimensions	Training Time (s)	Accuracy (%)	Macro-F1 (%)
Relu	256	402.84	82.59	49.86
Gelu	256	405.33	84.52	51.37
Relu	512	419.35	84.82	51.36
Gelu	512	417.25	83.87	51.01

https://doi.org/10.3390/mi1701006

Table 7. The ablation study for different hidden layers.

Hidden Layer	Training Time (s)	Accuracy (%)	Macro-F1 (%)
1	394.37	83.59	50.54
2	419.35	84.82	51.36
3	429.29	84.05	51.39

YOLO Backbone: To explore the performance of YOLOs, we compared the performance of different YOLOs architectures. The detailed results are shown in Tables 8–11. There are clear differences in terms of design objectives, computational efficiency, and accuracy. YOLOv8 showed a steady scaling trend in parameters with training time increasing from 6.9 h to 58 h and inference latency from 2.0 ms to 29.2 ms. While providing balanced performance, its accuracy remains relatively stable between 0.833 and 0.839, showing limited improvement with model size, indicating that for topology-driven task, excessive model capacity may amplify sensitivity to noise and local irregularities. These observations highlight that semantic prototype alignment acts as a form of semantic regularization, the benefit of which depends on the balance between visual representation capacity and task characteristics. YOLOv10 showed high accuracy enhancement, achieving the highest accuracy among the four YOLO versions, with up to 0.853 in the YOLO with s size. However, this improvement comes with the downside of significantly deeper networks, longer training times, and slower inference. Therefore, YOLOv10 is more suitable for scenarios where maximum accuracy is required and computational resources are sufficient. YOLOv11 shows a more balanced trade-off between accuracy and efficiency. With Top-1 accuracy in the range of 0.84–0.851, it remains competitive with YOLOv10 while reducing both training and inference overheads. YOLOv12 represents a breakthrough in computational efficiency. It maintains a compact parameter size while dramatically reducing the training time and inference latency. In summary, YOLOv10 achieved superior accuracy at the expense of computational cost, YOLOv11 achieved a more practical balance between accuracy and efficiency, and YOLOv12 offered a substantial leap in training and inference efficiency with minimal accuracy loss, making it particularly attractive for real-time and resource-constrained applications.

Table 8. The performance of different YOLOv8 architectures.

YOLOv8 Size	N	S	M	L	X
Params (M)	1.4	5.1	15.8	36.2	56.1
Training Time (h)	6.9	10.7	31.8	40.4	58.0
Inference Time (ms)	2.0	4.3	9.7	20.3	29.2
Layers	73	73	103	133	133
Accuracy	0.839	0.835	0.836	0.835	0.833

Table 9. The performance of different YOLOv10 architectures.

YOLOv10 Size	N	S	M	L	X
Params (M)	0.9	2.9	8.3	20.2	30.1
Training Time (h)	7.4	13.9	28.9	55.5	73.15
Inference Time (ms)	2.8	6.1	13.4	28.1	36.5
Layers	89	126	185	244	244
Accuracy	0.844	0.853	0.845	0.852	0.85

https://doi.org/10.3390/mi17010067

Table 10. The performance of different YOLOv11 architectures.

YOLOv11 Size	N	S	M	L	X
Params (M)	1.5	5.4	10.4	12.8	28.3
Training Time (h)	8.2	13.0	27.4	32.1	54.2
Inference Time (ms)	2.4	5.3	14.1	15.5	28.9
Layers	112	112	138	227	227
Accuracy	0.84	0.848	0.844	0.85	0.851

Table 11. The performance of different YOLOv12 architectures.

YOLOv12 Size	N	S	M	L	X
Params (M)	1.7	6.1	12.0	15.4	34.0
Training Time (h)	1.5	1.7	2.2	3.256	9.25
Inference Time (ms)	0.3	0.4	0.6	0.7	1.2
Layers	94	94	104	190	190
Accuracy	0.842	0.837	0.844	0.845	0.843

5. Conclusions

In this paper, we propose YOLO-LA, a lightweight prototype-based vision–language alignment framework that integrates a frozen YOLO classifier backbone with a frozen text encoder for wafer pattern detection. Semantic text prototypes are generated from natural-language defect descriptions using a text encoder, enabling interpretable classification via cosine similarity in a shared embedding space. In addition, an efficient learnable projection head is designed to achieve minimal extra computation. Extensive experiments on wafer map defect classification demonstrate that semantic alignment consistently improves classification robustness, particularly under low-resolution conditions where purely visual cues are insufficient to distinguish structurally similar defect patterns. The results further show that the proposed framework is robust to prompt formulation, benefits from a frozen visual backbone, and achieves an effective performance–efficiency trade-off with a lightweight projection head. In future work, we will explore few-shot and zero-shot defect recognition through dynamic prototype updating, as well as validation on real-world production data from semiconductor fabrication lines.

Author Contributions: Conceptualization, Z.W. and R.W.; methodology, Z.W.; software, R.W.; validation, J.C. and Y.Z.; formal analysis, Z.S.; investigation, Y.Y.; resources, Z.S.; data curation, R.W.; writing—original draft preparation, Z.W.; writing—review and editing, W.F.; visualization, R.W.; supervision, W.F.; project administration, J.C.; funding acquisition, Z.W. and J.C. All authors have read and agreed to the published version of the manuscript.

Funding: The study is supported by the National Natural Science Foundation of China (Grant No. 308 52205522) and the project of Doctoral Research Initiation Fund of Hubei University of Technology (Grant No. XJ2024001101) and the Hubei Key Laboratory of Modern Manufacturing Quality Engineering (Grant No. KFJJ-2025009).

Data Availability Statement: The original contributions presented in the study are included in the article; further inquiries can be directed to the corresponding author.

Conflicts of Interest: The authors declare no conflicts of interest.

https://doi.org/10.3390/mi17010067

Appendix A

Table A1. Textual detailed description of defects.

Defect Type	Description
Center	Most defective dies are densely concentrated around the wafer center (central cluster).
Donut	Defects form a donut pattern: a ring-like distribution with a relatively cleaner center.
Edge_Ring	Defects are distributed along the wafer perimeter, forming a ring-like band near the edge.
Edge_Loc	Defects form a localized cluster near the wafer boundary/edge, not forming a full ring.
Loc	Defects are confined to a small local region (single compact cluster) inside the wafer.
Random	Defects are scattered randomly across the wafer with no clear spatial structure.
Scratch	Defects align into a long, thin, linear streak crossing the wafer (scratch-like line).
Near_full	Defects cover most of the wafer surface area (nearly full wafer affected).

Table A2. Textual concise description of defects.

Defect Type	Description
Center	Center cluster
Donut	Donut ring
Edge_Ring	Edge ring band
Edge_Loc	Edge local cluster
Loc	Local cluster
Random	Random scatter
Scratch	Linear scratch
Near_full	Nearly full coverage

References

1. Wang, R.; Cheung, C.F.; Zang, Y.; Wang, C.; Liu, C. Material removal rate optimization with bayesian optimized differential evolution based on deep learning in robotic polishing. *J. Manuf. Syst.* **2025**, *78*, 178–186. [CrossRef]
2. Liu, C.; Zhu, Z.; Chen, J.; Wang, R. Effect of grain size on nano-scratching of RB-SiC via molecular dynamics simulation. *Vacuum* **2026**, *245*, 115001. [CrossRef]
3. Liu, L.; Ke, J.; Yin, T.; Yip, W.S.; Zhang, J.; To, S.; Xu, J. Cutting mechanism of reaction-bonded silicon carbide in laser-assisted ultra-precision machining. *Int. J. Mach. Tools Manuf.* **2024**, *203*, 104219. [CrossRef]
4. Chen, X.; Chu, J.; Zhu, Z.; Liu, C. Effect of temperature on surface morphology of single-crystal silicon in nanometric cutting. *Appl. Surf. Sci.* **2025**, *684*, 161957. [CrossRef]
5. Liu, C.; Yip, W.S.; Chen, J.; Wang, R.; Xu, J.; To, S. Atomic simulation of the temperature effect on fabrication mechanism of micro-structured surface on single-crystal silicon. *J. Manuf. Process.* **2025**, *133*, 238–248. [CrossRef]
6. Wang, Y.; Ni, D.; Huang, Z.; Chen, P. A self-supervised learning framework based on masked autoencoder for complex wafer bin map classification. *Expert Syst. Appl.* **2024**, *247*, 123601. [CrossRef]
7. Kim, D.; Pilsung, K. Dynamic clustering for wafer map patterns using self-supervised learning on convolutional autoencoders. *IEEE Trans. Semiconduct. Manufact.* **2021**, *34*, 444–454. [CrossRef]
8. Redmon, J.; Divvala, S.; Girshick, R.; Farhadi, A. You only look once: Unified, real-time object detection. In Proceedings of the IEEE Conference on Computer Vision and Pattern Recognition, Las Vegas, NV, USA, 27–30 June 2016.
9. Liu, W.; Anguelov, D.; Erhan, D.; Szegedy, C.; Reed, S.; Fu, C.Y.; Berg, A.C. Ssd: Single shot multibox detector. In Proceedings of the Computer Vision–ECCV 2016: 14th European Conference, Amsterdam, The Netherlands, 11–14 October 2016.
10. Lin, T.Y.; Goyal, P.; Girshick, R.; He, K.; Dollár, P. Focal loss for dense object detection. In Proceedings of the IEEE International Conference on Computer Vision, Venice, Italy, 22–29 October 2017; pp. 2980–2988.

https://doi.org/10.3390/mi17010067

11. Girshick, R. Fast r-cnn. In Proceedings of the IEEE International Conference on Computer Vision, Santiago, Chile, 7–13 Decemb 2015.
12. Wang, J.; Kelvin, C.K.; Loy, C.C. Exploring clip for assessing the look and feel of images. In Proceedings of the AAAI Conferen on Artificial Intelligence, Washington, DC, USA, 7–14 February 2023; Volume 37.
13. Gu, Z.; Zhu, B.; Zhu, G.; Chen, Y.; Tang, M.; Wang, J. Anomalygpt: Detecting industrial anomalies using large vision-langua models. In Proceedings of the AAAI Conference on Artificial Intelligence, Vancouver, BC, Canada, 20–27 February 202 Volume 38.
14. Qian, K.; Sun, T.; Wang, W. Exploring Large Vision-Language Models for Robust and Efficient Industrial Anomaly Detectic *arXiv* **2024**, arXiv:2412.00890. [CrossRef]
15. Cao, Y.; Xu, X.; Cheng, Y.; Sun, C.; Du, Z.; Gao, L.; Shen, W. Personalizing vision-language models with hybrid prompts f zero-shot anomaly detection. *IEEE Trans. Cybern.* **2025**, *55*, 1917–1929. [CrossRef] [PubMed]
16. Jocher, G.; Chaurasia, A.; Qiu, J. YOLO by Ultralytics. 2023. Available online: https://github.com/ultralytics/ultralyti (accessed on 1 December 2025).
17. Wang, A.; Chen, H.; Liu, L.; Chen, K.; Lin, Z.; Han, J.; Ding, G. Yolov10: Real-time end-to-end object detection. *arXiv* **202** arXiv:2405.14458.
18. An, J.; Putro, M.D.; Jo, K.H. Efficient residual bottleneck for object detection on cpu. In Proceedings of the 2022 Internation Workshop on Intelligent Systems (IWIS), Ulsan, Republic of Korea, 17–19 August 2022; pp. 1–4.
19. Khanam, R.; Hussain, M. Yolov11: An overview of the key architectural enhancements. *arXiv* **2024**, arXiv:2410.17725. [CrossRe
20. Tian, Y.; Ye, Q.; Doermann, D. Yolov12: Attention-centric real-time object detectors. *arXiv* **2025**, arXiv:2502.12524.
21. Wang, C.Y.; Liao, H.Y.M.; Yeh, I.H. Designing network design strategies through gradient path analysis. *arXiv* **202** arXiv:2211.04800. [CrossRef]
22. Sundaresan Geetha, A.; Alif, M.A.R.; Hussain, M.; Allen, P. Comparative Analysis of YOLOv8 and YOLOv10 in Vehicle Detectic Performance Metrics and Model Efficacy. *Vehicles* **2024**, *6*, 1364–1382. [CrossRef]
23. Wu, M.J.; Jang, J.S.R.; Chen, J.L. Wafer map failure pattern recognition and similarity ranking for large-scale data sets. *IEEE Trar Semiconduct. Manufact.* **2014**, *28*, 1–12.
24. Wang, W.; Wei, F.; Dong, L.; Bao, H.; Yang, N.; Zhou, M. Minilm: Deep self-attention distillation for task-agnostic compression pre-trained transformers. *Adv. Neural Inf. Process. Syst.* **2020**, *33*, 5776–5788.
25. Song, K.; Tan, X.; Qin, T.; Lu, J.; Liu, T.Y. Mpnet: Masked and permuted pre-training for language understanding. *Adv. Neural I Process. Syst.* **2020**, *33*, 16857–16867.
26. Liu, Y.; Ott, M.; Goyal, N.; Du, J.; Joshi, M.; Chen, D.; Stoyanov, V. Roberta: A robustly optimized bert pretraining approach. *arX* **2019**, arXiv:1907.11692.4.
27. Wang, L.; Yang, N.; Huang, X.; Jiao, B.; Yang, L.; Jiang, D.; Wei, F. Text embeddings by weakly-supervised contrastive pre-trainin *arXiv* **2022**, arXiv:2212.03533.

https://doi.org/10.3390/mi1701006

micromachines

Article

A Transferable Digital Twin-Driven Process Design Framework for High-Performance Multi-Jet Polishing

Honglei Mo [1], Xie Chen [1], Lingxi Guo [1], Zili Zhang [2], Xiao Chen [3], Jianning Chu [3] and Ruoxin Wang [4,*]

1 Shanghai Aerospace Control Technology Institute, Shanghai 201109, China; mohonglei1987@sina.com (H.M.); spwanx@163.com (X.C.); guolingxiz@yeah.net (L.G.)
2 State Key Laboratory of Ultra-Precision Machining Technology, Department of Industrial and Systems Engineering, The Hong Kong Polytechnic University, Hong Kong 999077, China; zili1.zhang@polyu.edu.hk
3 Hubei Key Laboratory of Modern Manufacturing Quality Engineering, School of Mechanical Engineering, Hubei University of Technology, Wuhan 430068, China; chenxiao1987jz@163.com (X.C.); jianning_chu@163.com (J.C.)
4 Department of Mechanical Engineering, University of Maryland College Park, College Park, MD 20742, USA
* Correspondence: rwang135@umd.edu

Abstract

The multi-jet polishing process (MJP) demonstrates high shape accuracy and surface quality in the machining of nonlinear and complex surfaces, and it achieves precise and adjustable material removal rates through computer control. However, there are still challenges in terms of machining efficiency, system complexity, and stability. In particular, maintaining the polishing quality presents a greater challenge when working conditions change. To overcome these issues, this paper conceptually proposes a digital twin (DT)-driven, human-centric design framework that integrates key factors of MJP, such as jet kinetic energy, nozzle structure, abrasive type, and machining path. Within this framework, a feature-encoded transfer learning-based model is introduced to enhance surface roughness prediction accuracy and robustness under varying working conditions. The effectiveness of the proposed model was verified by conducting experiments on 3D printed workpieces under two different MJP working conditions. The results show that our proposed method yields better predictive performance and cross-condition adaptability. Overall, this work provides a predictive modeling component that supports DT-driven process design, offering a practical and extensible perspective for optimizing complex ultra-precision manufacturing processes under data-scarce and uncertainty-dominated conditions.

Keywords: digital twin; surface roughness prediction; fluid jet polishing; transfer learning; machining process

1. Introduction

Fluid jet polishing (FJP) is an ultra-precision manufacturing process that involves mixing water with abrasive particles and ejecting this mixture through a small nozzle at high velocities, thereby achieving high surface quality. Different to the contact machining methods such as diamond cutting [1,2] and grinding [3], the non-contact finishing technique is favored for its ability to achieve high-precision polishing without causing overheating of the workpiece. In addition, FJP also avoids the issue of tool wear and is capable of adapting to the processing requirements of complex nonlinear surfaces. With computer control, FJP can achieve accurate processing effects and controllable material removal rates, making it suitable for processing hard alloys, ceramics, glass, quartz, and other special materials.

Due to its high-precision machining capabilities, FJP can handle complex freeform surfaces, thus becoming a dominant technology in the field of ultra-precision machining.

Despite the great potential of FJP technology in the field of precision machining, it still faces challenges and limitations, such as the need for improved processing efficiency, system complexity, and stability requirements. Moreover, most FJP systems use open-loop control systems that cannot effectively provide feedback during the polishing process to adapt to uncertainties during machining. Recently, with the development of intelligent manufacturing, many studies are emerging to address these issues [4,5]. Data-driven approaches have shown significant promise in predicting complex machining outcomes where physics-based models struggle. For instance, the recent precision manufacturing domain demonstrated the generalizing ability of convolutional neural network models in predicting workpiece profiles in electrochemical machining under varying pulse conditions [6]. Furthermore, recent advancements have focused on overcoming the "black-box" nature of these models. Wu et al. [7] introduced Explainable AI (XAI) frameworks to interpret model decisions in real-time cavity profile prediction, bridging the gap between data-driven predictions and physical process dynamics. In particular, digital twin (DT) technology has been widely applied in various fields, including product lifecycle management, processing, assembly processes, and predictive and health management. In the FJP process, DT technology has the potential to enhance the accuracy and efficiency of process simulation, optimization, and control by integrating real-time sensor feedback into simulation models and data analysis processes. Therefore, in-depth research into the application of DT technology in FJP, especially in terms of process parameters, material removal function models, and processing path algorithms, will be key to improving machining precision and efficiency.

In this paper, a DT-based method is proposed to optimize the multi-jet process (MJP), a more efficient FJP process. Firstly, recent research on DT in the field of machining is reviewed to identify the shortcomings of existing methods. The key factors of MJP, such as jet kinetic energy, nozzle design, abrasive type, processing path, and removal rate model, are analyzed. Subsequently, a concept of a DT-based human-centric interaction design framework for the MJP process is proposed. In this framework, a feature encoding-based transfer learning method is designed to realize efficient, precise, and stable polishing process quality prediction under various working conditions. The effectiveness of the proposed method is validated through polishing experiments on the 3D printed material.

2. Related Works

2.1. Digital Twin in Machining

As the manufacturing industry transforms towards intelligent manufacturing models, DT-driven processing systems stand out for their capabilities in sensing, predicting, decision-making, and controlling. These systems have three characteristics: (1) the integration of various DT models, (2) the systematic operations that are constituted by interactions between these models, and (3) the provisioning of diverse services through the combination of these operational processes. The analysis of DT-driven machining can be approached from three perspectives: component (point), connectivity (line), and loop (face).

(a) Component Perspective: DT models serve as high-fidelity virtual counterparts to physical entities. They are influenced by differing manufacturing and process requirements and typically include geometric models, knowledge models, and algorithmic models [8]. Geometric models represent the changing geometric states and physical alterations of an entity [9]. Knowledge models uncover complex data correlations within the machining process and support various functionalities like algorithm development and process opti-

https://doi.org/10.3390/mi17020226

mization [10]. Algorithmic models are crucial for forecasting the actual production process and enabling dynamic and real-time modifications to the machining process [11].

(b) Connectivity Perspective: Examining the flow and transformation of data unveils that DT machining systems embody three critical operations: from physical to digital (sensing), virtual analysis (cognition), and from digital back to physical (feedback and interaction). Sensing involves the real-time acquisition of physical data to support decision-making [12]; cognition is the process of acquiring knowledge or predictive information [13]; feedback generates or displays actions that alter the physical world [14]; and interaction encompasses human participation within a looped environment [15].

(c) Loop Perspective: The operation of a DT machining system is cyclical, providing looped services tailored to various manufacturing stages, including process planning, autonomous decision control, human-in-the-loop control, and quality inspection. Process planning integrates various factors and verifies simulation results through actual machining to improve quality [16]. Autonomous decision control adjusts and fine-tunes the machining process based on the sensed information [17]. Human-in-the-loop control refers to intelligent decision-making by on-site personnel with the assistance of human–machine interactions [18]. DT-driven quality inspection involves analyzing production data to monitor product quality [19].

In summary, DT technology, with its comprehensive models, operations, and services, offers an integrated and dynamically optimized environment for intelligent manufacturing, enhancing processing efficiency, quality, and flexibility. This advancement is unlocking transformative potential in the manufacturing sector. A focus on the loop perspective, particularly the process design phase and related decision models within the DT machining system, will be crucial in future explorations.

2.2. *Digital Twin-Driven Process Design*

The main aim of intelligent manufacturing is to enhance manufacturing performance. Consequently, post-designing a product that conforms to aesthetic standards, it is crucial to conduct manufacturing simulations to ascertain if the product's performance aligns with the anticipated benchmarks. Furthermore, it is imperative to evaluate the machining process to establish whether any modifications are necessary. Therefore, process design is crucial and important. The scope of research in process design spans products, processes, and systems. Utilizing iterative design outcomes during the process design phase can bolster the quality of machining. Product design is largely concerned with devising product functionality and the allocation of product tolerances [20]. In addition, process design is inherently intricate, demanding a holistic assessment of diverse elements, including the workpiece, machining equipment, and parameters [21].

For the design of the manufacturing process, DT plays a pivotal role in offering services such as mapping out manufacturing routes [22], optimizing parameters [23], designing eco-friendly processes [24], reapplying knowledge [25], and evaluating procedures [26]. In system analysis, researchers have delved into methods for appraising spindle performance, strategies for optimizing processes in response to fluctuating operational conditions and assessing the system's compatibility during the design phase [27]. These design instances, facilitated by iterative features, have demonstrated that DT-led design paradigms can significantly enhance the performance of products, processes, and systems.

2.3. *Research Gaps and Motivations*

Although many studies have explored DT-driven machining processes, several research gaps remain in the context of ultra-precision machining:

https://doi.org/10.3390/mi17020226

(1) While DT technology has been recognized as an important technique to enhance the intelligence of manufacturing systems, design-level DT-driven frameworks for ultra-precision machining remain underexplored.

(2) Many studies on the method of machining process design already exist [28], but most of these methods are severely constrained by the original data collection scenarios. Therefore, their predictive performance and applicability often degrade significantly when process conditions change. Currently, there is a lack of effective modeling strategies, making it impossible to reuse previous knowledge and adapt to new processes or working conditions.

Therefore, in this paper, we propose a novel human-centric process design framework and a cross-domain adaptation mechanism to fill the research gaps mentioned above.

3. Digital Twin-Driven Process Design Method for Multi-Jet Polishing

3.1. Multi-Jet Polishing Process

MJP is a type of FJP, which utilizes multiple nozzles to improve polishing efficiency and uniformity. The fundamental mechanism of MJP is illustrated in Figure 1. It adopts multiple pressurized polishing slurry jets to improve the surface finishing where abrasive particles are accelerated to impinge on the workpiece surface, leading to nano-scale material removal. Basically, the surface generation of MJP is determined by both the Tool Influence Function (TIF) and the tool path planning. The TIF is assessed in terms of maximum depth, width and volumetric material removal rate, and can be affected by various polishing parameters including the abrasive particle concentration, slurry pressure, nozzle geometries, tool offset distance, impingement angle, etc. Two main factors that affected the impinging velocity and impact force of abrasive particles on the target surface were examined which are the fluid pressure (P) and tool offset distance (TO). Theoretically, higher pressure and smaller tool offset distance result in deeper material removal. Despite the TIF, tool path planning is also a key factor to the surface generation in MJP. The surface roughness profile of the polished surface is formed by the continuous repetitive impact of abrasive particles along the raster path under a specific tool feed rate (f) and step distance (d). The feed rate had controlled the dwelling time for material removal while the step distance controlled the overlapping area of the removal profiles. The detailed illustration is shown in Figure 1. The parameters of MJP can be outlined as follows:

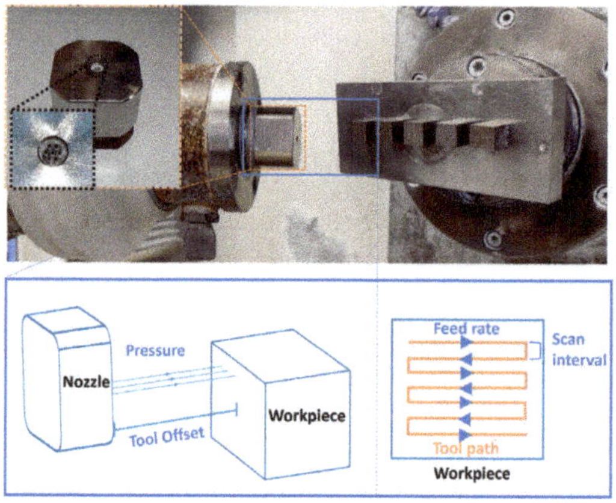

Figure 1. Mechanism of MJP.

https://doi.org/10.3390/mi1702022

Kinetic Energy Parameters: The kinetic energy of the jet is principally manifested through the energy and velocity of the abrasive particles. The greater the energy and velocity imparted to these particles are, the more significant their effect on the workpiece's surface is. To augment the efficiency of material removal and the precision of the machining process, a thorough investigation is imperative for the sensitive parameters that influence the kinetic energy of the abrasives, such as the jet's pressure, angle, standoff distance, and abrasive characteristics.

Nozzle Design: The design, quantity, and motion pattern of the nozzles in MJP play a substantial role in the machining performance. A qualitative analysis of the nozzle performance is necessary for achieving high surface quality.

Abrasives: The factor of the abrasives, such as shape, size, and type, have a direct impact on the efficiency of the machining process and the surface quality.

3.2. The Framework for Digital Twin-Driven Process Design Method

The dynamic nature of machining environments, where processes evolve over time, often leads to a decrease in the efficiency of process planning execution. Before implementing a process, it is crucial to evaluate whether the current process content meets the machining requirements of the workpiece and can produce a finished product of specified quality. Furthermore, it is essential to perform optimization calculations on real-time data and critical parameters during the process implementation, to provide a preselection of the optimal parameters and effective numerical references. Therefore, a human-centric, DT-driven process design framework is conceptually introduced, as illustrated in Figure 2. In the proposed framework, process data collected from physical polishing experiments, including process parameters and measured surface roughness, are used to update a virtual predictive model at the process-design level. Based on the updated model, predicted quality metrics and candidate parameter suggestions can be generated to support subsequent process planning and decision-making. Human designers remain in the loop and interpret these predictions when selecting or adjusting process parameters. In the proposed human-centric design approach, the DT model serves as the sole data source during the three-dimensional process design, playing a vital role in conveying process information. The core of the DT-driven human-centric process design method is the generation of an interactive process model based on the design model. The process design enables the creation of a three-dimensional solid process model that can manage process information. The automatic extraction of design model information provides a foundational layer of information. The construction of a process knowledge base, consisting of process instance libraries, process rule libraries, and equipment resource libraries, supports the reuse of process instances and automatic process decision-making within the aid of process design systems. Automatic process decision-making includes selecting machining methods, equipment, planning process routes, computer numerical control (CNC) machining operations, and computing decisions to achieve knowledge-based automatic planning of process routes and programmatic calculation of process parameters, generating comprehensive and detailed process information. Finally, the generated process information is integrated into a process information tree. Combined with model lightweight processing, a three-dimensional operation manual is created to realize visual guidance on the shop floor.

Based on the DT-driven process design approach in Figure 2, experiments are conducted in an iterative manner. First, an initial set of feasible process parameters is selected based on machine constraints and expert experience. Second, experiments are conducted on the physical system under the selected settings. Third, the surface roughness is measured after polishing and used as quality feedback. Fourth, the collected process quality data are incorporated to update the DT-driven prediction model. Finally, the updated model

https://doi.org/10.3390/mi17020226

provides quality predictions under candidate parameter settings, which supports selecting the next trial settings for subsequent experiments.

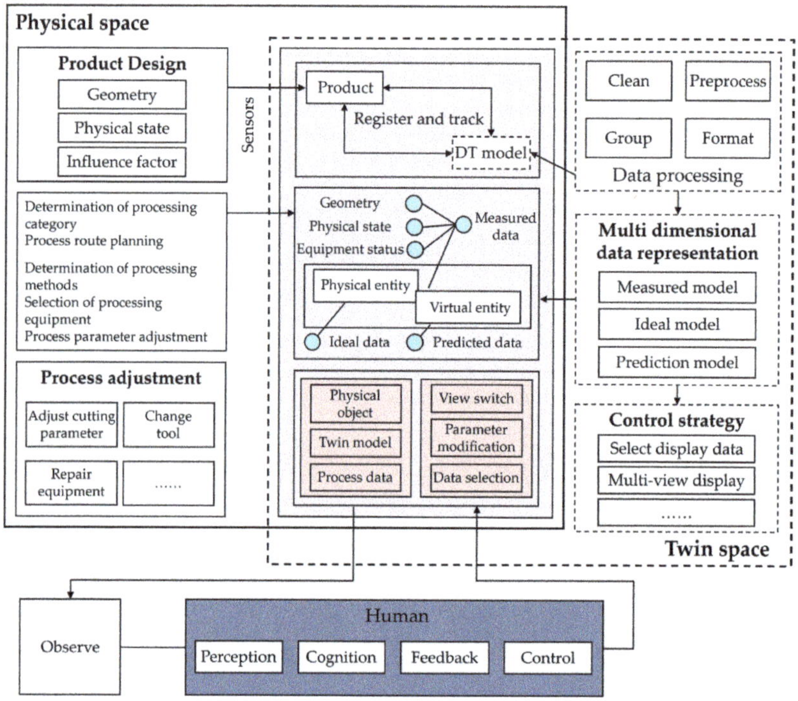

Figure 2. Conceptual architecture of a human-centric, DT-driven process design framework. The DT is to operate at the process-design level, where experimental data update the virtual model and predicted quality metrics provide feedback for process parameter optimization in a human-in-the-loop manner.

Therefore, compared to traditional process evaluation methods, the human-centric process design approach has the following characteristics:

Multi-Modal Process Information Perception: The perception of the machining process relies on the equipment configured on-site to collect data. However, the fidelity of this perception depends heavily on understanding the process context (e.g., sensor positioning and signal source motion). As demonstrated by Wu et al. [29] in laser powder bed fusion, acoustic signals can vary significantly based on the relative position of the emission source, necessitating data-driven methods to identify these contextual factors for accurate monitoring. Consequently, the resulting massive twin dataset consists of multi-source heterogeneous data mappings that must be contextually aligned before being transmitted to the digital twin, enabling the virtual model to accurately reflect the physical process state. These data are used to continuously update the virtual process model, enabling synchronization between physical experiments and the DT at the process design level. The human-centric process design model supports these twin data, achieving optimization through machining quality prediction and refining machining paths, process methods, and process parameters.

Quality-Centric Approach: The entire process is quality-driven, with continuous adjustments made based on quality metrics derived from the virtual simulations and real-

https://doi.org/10.3390/mi17020226

time data analysis, ensuring that the final product meets the specified standards. In this manner, predicted quality indicators serve as feedback from the DT to guide process design.

Dynamic Adaptation: The combination of real-time data with simulation allows for dynamic adaptation of the process, ensuring that any deviations from the desired outcomes can be corrected promptly, leading to a more agile and resilient manufacturing process. This adaptability is further enhanced by transferring knowledge learned from previous working conditions.

Proactive Problem-Solving: By placing humans in the loop, the design method allows for preemptive actions to be taken, ensuring that any potential issues can be anticipated and mitigated effectively, thus reducing downtime and enhancing the overall efficiency of the process.

Human-In-The-Loop Design Mode: The system uses virtual simulation technology to explore and predict the unknown world, followed by process modifications based on quality information until process execution is completed. Process designers can take proactive measures to avoid or solve potential disturbances in advance. This forms a closed decision-making loop between the physical process, the DT, and human designers.

3.3. The Adaptability of Process Design

In manufacturing industry, the adaptability of process design is crucial for ensuring product performance and manufacturing efficiency. Current research indicates that due to the complexity of manufacturing environments and the variability of process parameters, static process design models struggle to meet the rapidly changing market demands. Fluctuations in the environment, changes in material properties, and differences in operator skills are key factors that affect the stability of process design and product performance. Moreover, to meet the demands for personalized and customized products, process design must be flexible to accommodate continuously updated product characteristics and manufacturing technologies.

Adaptive process design models can adjust process parameters in real time, ensuring that product performance meets predefined standards despite fluctuations in raw material supply, changes in environmental conditions, and variations in production equipment performance. These models, through real-time data collection, advanced the data processing technology, and machine learning algorithms can monitor and predict the impact of key process parameters on product performance, making immediate adjustments when deviations are detected. In a production mode of multiple varieties and small batches, they can quickly switch between different products, reducing the time and cost to set new process parameters, thereby increasing production flexibility and efficiency. As such, adaptive process design models can learn and predict the process conditions required for new product designs, quickly adapt to changes on the production line, reduce downtime, and enhance the overall production capacity.

With the advancement of manufacturing technology, especially in Industry 4.0, the importance of adaptive process design models will be further enhanced to support more intelligent and automated production systems. Thus, the demand for adaptability in product quality prediction models is also growing. The core component of the adaptive process design model is the adaptability of the product quality prediction model. Transfer learning is an important pathway to improve the adaptability of product prediction models. In this work, we focus on representation-level adaptability, which motivates the use of feature-encoded transfer learning. This strategy allows invariant process representations to be learned from historical data while preserving flexibility for adaptations under new working conditions with minimal additional data.

https://doi.org/10.3390/mi17020226

4. Transfer Learning-Based Method

This section introduces a feature encoding-based transfer learning prediction model for predicting surface roughness, which takes into account the workpiece material (i.e., 316L stainless steel 3D printed parts), polishing parameters (feed rate (f), fluid pressure (P), tool offset (TO), and distance (d)), as well as surface roughness parameters (i.e., initial surface roughness, Ra_i, and final surface roughness, Ra_f). Figure 3 illustrates the framework of the proposed feature-encoded transfer learning (FETL) model. The model consists of two core modules: (a) the feature encoding module and (b) the regression prediction module. After encoding the input data in the feature encoding module, the generated feature values are passed to the regression prediction module, where the final predicted surface roughness is determined through the guidance aggregation of multiple sub-decision trees. Therefore, domain-invariant representations can be extracted to preserve essential process–quality relationships while suppressing task-specific variations to avoid performance degradation due to distribution shifts caused by changes in materials, nozzle configurations, and process parameters.

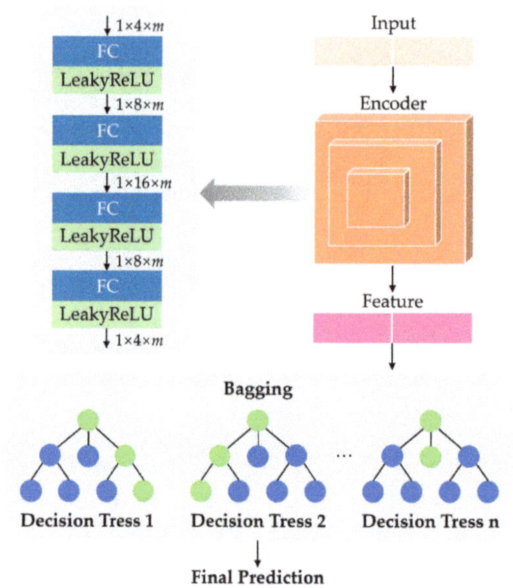

Figure 3. The architecture of proposed network.

4.1. Transfer Learning Strategy

During the process of polishing different workpiece materials, there is a shift in the data concerning polishing parameters and surface roughness. When the same model is used to perform both tasks simultaneously, this shift can lead to a significant reduction in model accuracy. To make full use of polishing data from different materials and processes, a feature encoding-based transfer learning strategy is proposed in this paper. This strategy enables the model to transfer from the labeled data of different tasks (source domain data) to the unlabeled data of the new task (target domain data), thus training a model suitable for the target domain. The core idea of this strategy is that an encoder is designed to process the input data and learn a domain-invariant feature representation. The encoder can be intuitively understood as a process-aware transformation that maps heterogeneous polishing conditions into a unified latent space, enabling the regression model to focus on the fundamental polishing mechanism. In addition, the effectiveness of the proposed

https://doi.org/10.3390/mi17020226

transfer strategy depends on the similarity of the underlying process physics between the source and target domains.

As shown in Figure 4, the training process of the prediction model is divided into two stages. In the first stage, the data from the source and target domains are randomly mixed and then input into the feature encoding module to learn a common representation of the data between domains. In the second stage, using the encoding module trained in the first stage, the feature representation of the current task's input data is obtained and passed into the regression prediction module, which learns the mapping between the encoded features and the predicted values. When training the encoding module, the loss function is designed to focus on the information loss between the encoded feature and the original input, guiding the learning direction of the encoding module and eliminating the impact of model overfitting on the target domain's feature recognition ability.

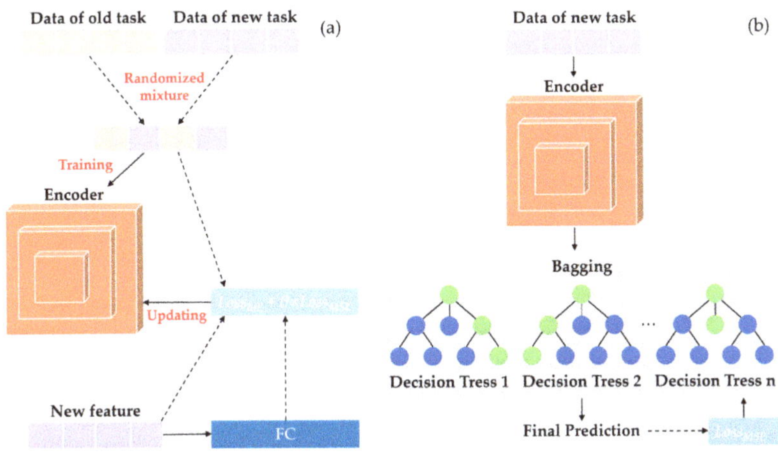

Figure 4. The training process of the prediction model in (**a**) stage 1: feature encoder training using mixed source and target domain data to learn shared representations; and (**b**) stage 2: regression model training based on encoded features for surface roughness prediction.

The training of the encoder module consists of forward propagation and backward parameter tuning processes. Assuming that the network processes m training samples $\left\{\left(x^{(i)}, y^{(i)}\right)\right\}_{i=1}^{m}$ in one batch, where $x^{(i)} \epsilon \mathbb{R}^4$ is the input vector and $y^{(i)}$ is the labeled final surface roughness for that sample, then the forward propagation calculates the input features at each layer as follows:

$$x^l = f\left(W^l x^{l-1} + b^l\right) \quad (1)$$

where x^{l-1} and x^l denote the input and output feature vectors of the l-th layer, respectively; W^l and b^l are the weights and bias vectors of the neurons in the l-th layer network, respectively. $f(\cdot)$ is the activation function defined as follows:

$$\mathrm{LeakyReLU}(x) = \begin{cases} x, x \geq 0 \\ \eta x, x < 0 \end{cases} \quad (2)$$

https://doi.org/10.3390/mi17020226

where $\eta\epsilon(0, 1)$ is a predefined constant. For the training set containing m samples, the overall objective function of the encoder module is defined as follows:

$$J(W,b) = \frac{1}{m}\sum_{i=1}^{m}\frac{1}{2}\left\|h_{W,b}\left(x^{(i)}\right) - y^{(i)}\right\|^2 + \frac{\lambda}{2}\sum_{l=1}^{n_l-1}\sum_{i=1}^{s_l}\sum_{j=1}^{s_l+1}\left(W_{ij}^{(l)}\right)^2 \quad (3)$$

Here, λ stands for the weight attenuation parameter, n_l is the total number of layers in the network, s_l is the total number of neurons in the l-th layer of the network, $h_{W,b}\left(x^{(i)}\right)$ is the total output of the encoding module under the current weight W and bias b, and $W_{ji}^{(l)}$ is the weight of neurons in the l-th layer.

During backpropagation, the parameters are updated using gradient descent:

$$W_{ij}^{(l)} = W_{ij}^{(l)} - \alpha\frac{\partial}{\partial W_{ij}^{(l)}}J(W,b) \quad (4)$$

$$b_i^{(l)} = b_i^{(l)} - \alpha\frac{\partial}{\partial b_i^{(l)}}J(W,b) \quad (5)$$

where α is the learning speed.

When the cost error of the entire network structure is minimized, a pre-trained encoder module is obtained. The encoder module proposed in this paper consists of four basic blocks connected in series, each block containing a fully connected layer and an activation function layer. The input data are $1 \times 4 \times m$ vectors, and the first two blocks gradually expand to $1 \times 8 \times m$ and $1 \times 16 \times m$ vectors, respectively. Then, through the last two blocks, the vector is gradually compressed back to $1 \times 8 \times m$ and finally to a $1 \times 4 \times m$ feature vector. In the end, the encoding module outputs a feature vector whose dimensions are the same as the original input data. The encoder is optimized jointly with the surface roughness prediction objective and a cross-domain information constraint to learn task-relevant and domain-invariant feature representations.

The training of the regression prediction module consists of four parts: data preparation, decision tree construction, ensemble prediction, and parameter updating. First, the original input data are processed using the trained encoding module from stage one, which provides the input data for the regression prediction module. Second, a random training sample set is obtained using a bootstrapping sampling method, and fully split child decision trees are built. Then, the final predicted surface roughness is determined through the ensemble of multiple sub-decision trees. Finally, the predictions are compared with the actual values, and the parameters of the regression prediction module are updated accordingly.

4.2. Loss Function

Loss function for stage 1: The purpose of the first stage of training is to enable the encoding module to learn a feature expression that is invariant across domains from the mixed data of the target and source domains, in order to better accomplish the surface roughness prediction task. However, for this feature expression, it is not possible to obtain true label values to calculate the loss function. Therefore, this paper focuses on the error of the final prediction task, as well as the information loss between the feature and the original input data, designing the following loss function:

$$Loss_1 = Loss_{InE} + \theta Loss_{MSE} \quad (6)$$

https://doi.org/10.3390/mi17020226

where $Loss_{InE}$ is information loss, $Loss_{MSE}$ is the mean squared error loss, and θ is the balancing weight coefficient. The information loss is used to measure how much information is lost when the feature expression is used; expressing the similarity between the two is implemented using the Kullback–Leibler (KL) divergence as shown in Equation (7):

$$Loss_{InE} = \sum_{i=1}^{d} p_i ln \frac{p_i}{q_i} = D_{KL}(p \parallel q) \tag{7}$$

where d is the feature dimension and p_i and q_i represent the probabilities associated with the i-th feature component in the target and source domains, respectively. The corresponding probability distributions $p = \{p_i\}$ and $q = \{q_i\}$ are obtained by applying Softmax normalization to the feature vectors as shown in Equation (8):

$$p_i = \frac{\exp(f_i)}{\sum_{j=1}^{d} \exp(f_j)}, \; q_i = \frac{\exp(g_i)}{\sum_{j=1}^{d} \exp(g_j)} \tag{8}$$

where $f = \{f_i\}$ and $g = \{g_i\}$ denote the encoded feature vectors extracted from the target and source domains. The mean squared error (MSE) loss calculates the square difference between the predicted values and the target values, defined in Equation (9):

$$Loss_{MSE} = \frac{1}{n} \sum_{i=1}^{n} (y_i - \hat{y}_i)^2 \tag{9}$$

In the first stage of training, the input to the encoding module is passed to the fully connected layer, and its output is used as the final predicted surface roughness, which is then used to calculate the MSE loss with ground-truth. This is weighed up with the information loss calculated from the encoded features to obtain the overall total loss. The model parameters of each block of the encoding module are updated through backpropagation.

Loss function for stage 2: The purpose of the second stage of training is to enable the regression prediction module to learn the mapping between the encoded features and the predicted surface roughness. During the model training process, it is only necessary to focus on the error of the prediction task, and therefore the MSE loss function is used as the overall loss function for stage two.

5. Experimental Verification

5.1. Experimental Setup

During the MJP experiment, the ZEEKO IRP200 machine (Zeeko Limited, Coalville, UK) was used for polishing with different process parameters, as shown in Figure 5. Two different working conditions were setup, using seven nozzles and four nozzles, respectively (see Figure 5c,d). The polishing setup is shown in Figure 5b, with 3D printed 316L stainless steel components and CoCr (see Figure 5e,f). In addition, the explicit process parameters used by the predictive model are listed in Table 1. Process factors such as nozzle configuration and jet kinetic energy are treated as domain-level conditions, and their effects are incorporated through transfer learning across different domains. After polishing, the surface roughness (*Ra*) was measured three times and averaged to obtain the final Ra. For the two conditions, 43 and 80 sets of experimental datasets with different polishing parameters were collected, including feed rate (*f*), fluid pressure (*P*), tool offset (TO), step size (*d*), initial surface roughness (Ra_i), and surface direction. All experimental data were collected under the assumption that the sensing environment remained stable. The process parameters are treated as continuous or quasi-continuous variables, and the predictive model aims to learn an underlying process–quality mapping from the sparsely

https://doi.org/10.3390/mi17020226

sampled data. For the target domain, the dataset is first split into training and testing subsets. During Stage 1, the encoder is trained using a mixed dataset consisting of the source domain data and the training subset of the target domain only. The target domain test data are not used in any stage of encoder training. During stage 2, the regression model is trained using the encoded features from the target domain training set and evaluated exclusively on the held-out target domain test set. The dataset covers different materials, nozzle configurations, and initial surface states, which provides a basis for evaluating cross-condition adaptability.

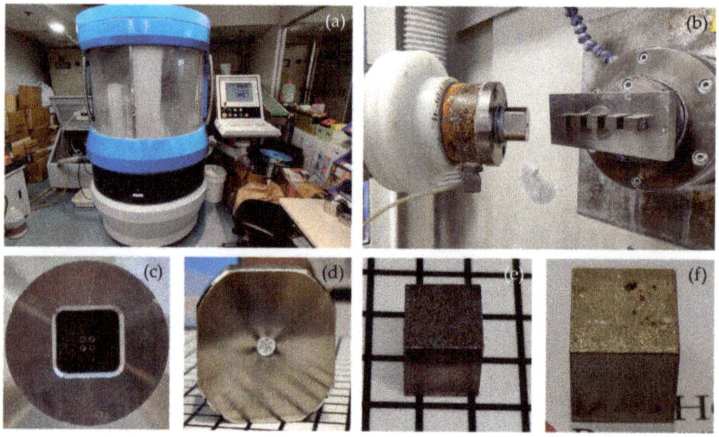

Figure 5. Experiment setup for multi-jet polishing. (**a**) The ZEEKO IRP 200 polishing equipment, (**b**) the polishing setting, (**c**) the 4-jet nozzle, (**d**) the 7-jet nozzle, (**e**) the 316L stainless steel workpiece, and (**f**) the CoCr workpiece.

Table 1. Explicit process parameters used as model inputs under different nozzle configuration domains.

Parameter	Working Condition 1	Working Condition 2
Injector	7	4
Material	316L stainless steel	CoCr
Feed rate size (mm/min)	10, 15, 20, 25, 30, 40, 60, 80	10, 15, 20, 25, 30, 40, 60, 80
Fluid pressure (bar)	4, 5, 6, 7, 8, 9, 10	4, 5, 6, 7, 8, 9, 10
Tool offset (mm)	2.5, 5, 7.5, 10, 12.5, 15	2.5, 5, 7.5, 10, 12.5, 15
Step size (mm)	0.1, 0.2, 0.3, 0.4, 0.5, 0.6, 0.7, 0.8	0.1, 0.2, 0.3, 0.4, 0.5, 0.6, 0.7, 0.8
Initial surface roughness (nm)	400~500	2000~4200
Surface direction	TS (Top)	TS (Top)

Our model was implemented in PyTorch and optimized using the Adam optimizer, with the parameters set to $\beta\_1 = 0.9$ and $\beta\_2 = 0.999$. The proposed model was compiled on a computer server equipped with a Tesla V100 PCIe 32 GB GPU card (NVIDIA Corporation, Santa Clara, CA, USA), using Python 3.6, Pytorch 1.8, and CUDA 10.2.

5.2. Surface Topography Analysis Before and After Multi-Jet Polishing

The surface quality of 316L stainless steel and CoCr after the 3D printing process were evaluated. As shown in Figure 6, a clear observation from the before-and-after comparison indicates that the laser melting traces produced during the 3D printing process were effectively eliminated by the MJP procedure. Moreover, surface defects such as un-melted particles and cracks were also significantly improved. However, based on the surface comparison scanning electron microscope (SEM) results before and after polishing, there are still some larger-scale wave-like textures present. This is because the micrometer-sized

https://doi.org/10.3390/mi17020226

particles in the abrasive media used during polishing can only remove roughness to a certain extent and are not effective for millimeter- or sub-millimeter-level irregularities.

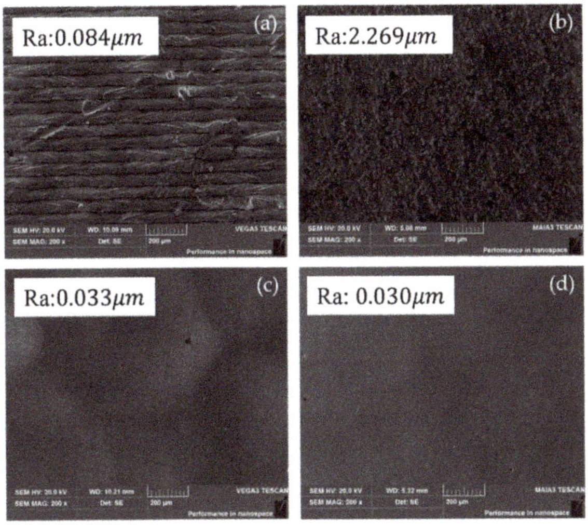

Figure 6. Scanning electron microscope images of the 316L workpiece before and after multi-jet polishing. (**a**) The top surface (Ra: 0.084 μm) and (**b**) side surface (Ra: 2.269 μm) of workpiece before polishing, respectively. (**c**) The top surface (Ra: 0.033 μm) and (**d**) side surface (Ra: 0.030 μm) of workpiece after polishing, respectively.

5.3. Comparison with State-of-the-Art Methods

Comparison experiments were conducted with state-of-the-art (SOTA) methods, including XGBoost, RF (Random Forest), LASSO (Least Absolute Shrinkage and Selection Operator), and SVR (Support Vector Regression). The numerical results of these methods are presented in Figure 7. Our method outperformed in all three metrics: MAE (Mean Absolute Error), MSE, and R2 (Coefficient of Determination). MAE, MSE, and R2 are metrics used to measure the predictive capability of regression models. They quantify the model's prediction accuracy by calculating the differences between the true values and the predicted values.

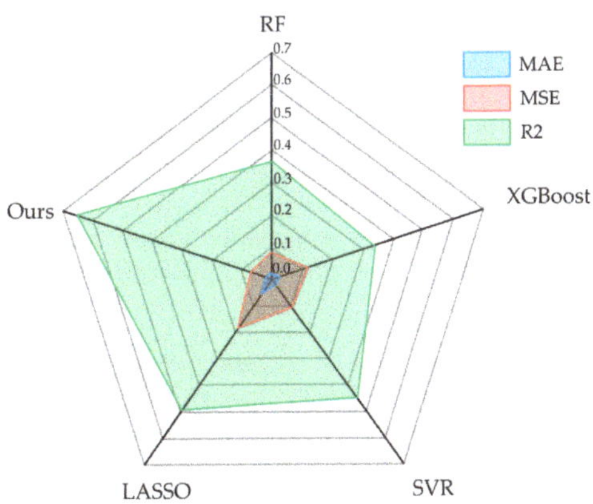

Figure 7. Comparison of prediction accuracy with SOTA model.

https://doi.org/10.3390/mi17020226

MAE is less sensitive to outliers, but it does not reflect the distribution of predictive errors. MSE is sensitive to outliers (because when the error is greater than 1, squaring it will further amplify the value), but it can reflect the distribution of predictive errors. R2 indicates the proportion of the variance for the dependent variable that is explained by the model; its value ranges from 0 to 1, with values closer to 1 indicating a better fit of the model to the data.

As shown in Figure 7, our method achieved the lowest MAE and MSE, indicating the smallest discrepancy with the true values. The highest MAE and MSE were observed for LASSO, indicating the lowest prediction accuracy. The other three methods had similar metric values, suggesting a comparable level of discrepancy between their predictions and the actual values. From the R2 metric perspective, our model had the best fit to the data, while XGBoost had the poorest fit. The results also indicate that the prediction errors calculated by MAE and MSE do not fully represent the model's fitting ability.

5.4. Ablation Experiments

In this section, an ablation study will be conducted to determine the effectiveness of the main innovations in the proposed method. Specifically, the impact of the loss function, transfer strategy, and feature transformation [30] on performance improvement was studied. The quantitative results of the ablation experiments are shown in Table 2. By learning shared feature representations, the model can effectively reuse information across materials and nozzle configurations, alleviating performance degradation caused by limited target domain samples.

Table 2. Ablation test.

Methods	MAE	MSE	R^2
Ours	0.066	0.007	0.643
Min–Max	0.073	0.007	0.616
w/o Transfer	0.081	0.011	0.453
MSE Loss	0.076	0.007	0.622

Compared to using only MSE loss, our method reduced MAE (Mean Absolute Error) by 0.01 and increased R2 by 0.021, with no change in MSE. Training with MSE loss directly targets the reduction in mean squared error, which is sensitive to outliers, so the model training results tend to eliminate outliers, which may lead to a homogenization of prediction results and not learning the true distribution of the data. Compared to not using a transfer strategy, R2 increased by 0.19, MAE decreased by 0.015, and MSE decreased by 0.04. The purpose of the transfer is to make full use of the existing data, applying the knowledge learned from similar data to new tasks, which is why there is a significant improvement in all three evaluation metrics. Compared to the common min–max feature transformation method, the proposed method has improvements in both MAE and R2, but the extent of improvement is smaller compared to others. Figure 8 shows the fit of different models to each sample in the ablation study, where Figure 8a reports point predictions obtained by averaging the outputs of the bagging ensemble of our proposed model.

https://doi.org/10.3390/mi17020226

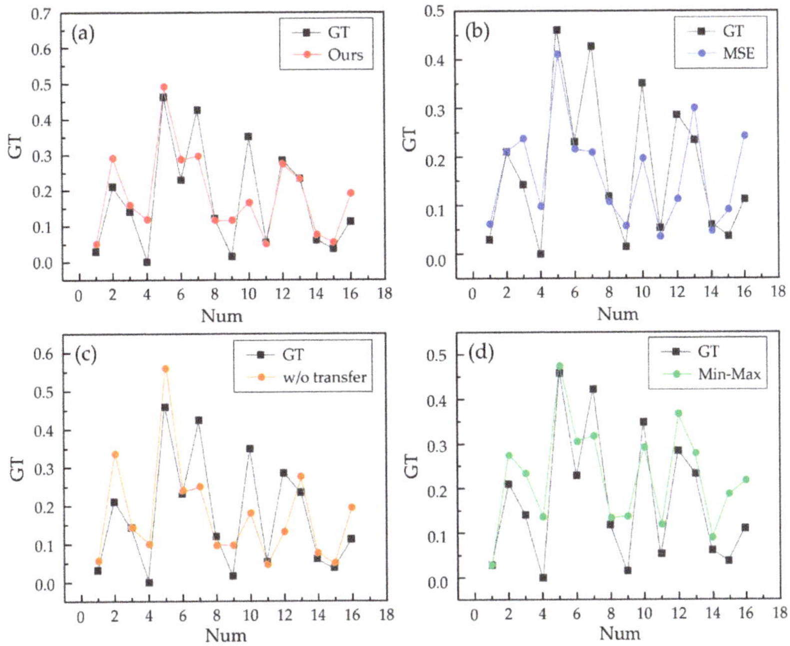

Figure 8. Results of ablation test. (**a**) Ours, (**b**) MSE loss, (**c**) w/o transfer, and (**d**) min–max.

6. Conclusions

In conclusion, this study presents a digital twin (DT)-driven process design framework for multi-jet polishing (MJP), aiming to address the challenges associated with complexity and uncertainty under changing working conditions. The proposed framework focuses on enhancing decision support during process design through improved quality prediction.

The human-centric design approach integrates critical factors of DT-driven MJP, including jet kinetic energy, nozzle structure, abrasive type, and machining path. In particular, the incorporation of a feature-encoded transfer learning strategy enables more accurate and robust surface roughness prediction across different polishing conditions by embedding standard encoding architectures into precision manufacturing workflows. The framework provides a generalizable approach for improving robustness and efficiency in data-scarce and uncertainty-dominated machining environments. The validation experiments on 3D printed workpieces were conducted to demonstrate the effectiveness.

Although the proposed transfer learning-based method demonstrates strong predictive performance, its applicability under increasing source–target discrepancies has not been systematically explored. In future work, we will evaluate the generality of the proposed framework in other ultra-precision manufacturing processes, such as bonnet polishing, chemical mechanical polishing, and magnetic abrasive finishing, as well as the integration of real-time system deployment [7], human–machine interaction interfaces, and trustworthy physical–virtual consistency under evolving sensing configurations.

Author Contributions: Conceptualization, investigation, and writing—original draft preparation, H.M.; methodology and software, H.M. and X.C. (Xie Chen); validation, Z.Z. and R.W.; formal analysis, X.C. (Xie Chen); data curation, L.G.; project administration, J.C.; funding acquisition, X.C. (Xiao Chen); supervision and writing—review and editing, Z.Z. and R.W. All authors have read and agreed to the published version of the manuscript.

https://doi.org/10.3390/mi17020226

Funding: The study is supported by the National Key R&D Program of China (Grant No. 2025YFB3411702) and the Hubei Key R&D Program (Grant No. 2023BAB195).

Institutional Review Board Statement: Not applicable.

Informed Consent Statement: Not applicable.

Data Availability Statement: The original contributions presented in this study are included in the article. Further inquiries can be directed at the corresponding author.

Conflicts of Interest: The authors declare no conflicts of interest.

Abbreviations

The following abbreviations are used in this manuscript:

DT	Digital Twin
RF	Random Forest
SVR	Support Vector Regression
MJP	Multi-Jet Polishing
MAE	Mean Absolute Error
FJP	Fluid Jet Polishing
UPM	Ultra-Precision Machining
LASSO	Least Absolute Shrinkage and Selection Operator
R2	Coefficient of Determination
MSE	Mean Square Error

References

1. Liu, C.; Ke, J.; Yin, T.; Yip, W.S.; Zhang, J.; To, S.; Xu, J. Cutting mechanism of reaction-bonded silicon carbide in laser-assiste ultra-precision machining. *Int. J. Mach. Tools Manuf.* **2024**, *203*, 104219. [CrossRef]
2. Du, H.; Zhao, D.; Liu, C.; Chen, H.; To, S. Laser-assisted slow tool servo diamond turning of single-crystal silicon for fabricatin micro-lens arrays. *J. Manuf. Process.* **2024**, *127*, 174–183. [CrossRef]
3. Liu, C.; Yip, W.S.; To, S.; Chen, B.; Xu, J. Numerical Investigation on the Effects of Grain Size and Grinding Depth on Nan Grinding of Cadmium Telluride Using Molecular Dynamics Simulation. *Nanomaterials* **2023**, *13*, 2670. [CrossRef] [PubMed]
4. Wang, R.; Cheung, C.F.; Zang, Y.; Wang, C.; Liu, C. Material removal rate optimization with bayesian optimized differenti evolution based on deep learning in robotic polishing. *J. Manuf. Syst.* **2025**, *78*, 178–186. [CrossRef]
5. Liu, S.; Bao, J.; Zheng, P. A review of digital twin-driven machining: From digitization to intellectualization. *J. Manuf. Syst.* **202** *67*, 361–378. [CrossRef]
6. Wu, M.; Arshad, M.H.; Saxena, K.K.; Qian, J.; Reynaerts, D. Profile prediction in ECM using machine learning. *Procedia CIR* **2022**, *113*, 410–416. [CrossRef]
7. Wu, M.; Yao, Z.; Verbeke, M.; Karsmakers, P.; Gorissen, B.; Reynaerts, D. Data-driven models with physical interpretability fc real-time cavity profile prediction in electrochemical machining processes. *Eng. Appl. Artif. Intell.* **2025**, *160*, 111807. [CrossRef
8. Zheng, P.; Sivabalan, A.S. A generic tri-model-based approach for product-level digital twin development in a smart manufactu ing environment. *Robot. Comput.-Integr. Manuf.* **2020**, *64*, 101958. [CrossRef]
9. Pashkov, A.E.; Malashchenko, A.Y.; Pashkov, A.A. On Creating Digital Technologies for the Production of Large Aircraft Fram and Skin Parts. *Russ. Metall. Met.* **2022**, *2021*, 1777–1785. [CrossRef]
10. Ladj, A.; Wang, Z.; Meski, O.; Belkadi, F.; Ritou, M.; Da Cunha, C. A knowledge-based Digital Shadow for machining industry i a Digital Twin perspective. *J. Manuf. Syst.* **2021**, *58*, 168–179. [CrossRef]
11. Giuliano, G.; Corrado, A.; Polini, W. A Geometric Algorithm to Evaluate the Thickness Distribution of Stretched Sheets throug Finite Element Analysis. *Appl. Sci.* **2021**, *11*, 1905. [CrossRef]
12. Ganser, P.; Venek, T.; Rudel, V.; Bergs, T. Dpart—A Digital Twin Framework for the Machining Domain. *MM Sci. J.* **2021**, *202* 5134–5141. [CrossRef]
13. Xia, L.; Zheng, P.; Li, X.; Gao, R.X.; Wang, L. Toward cognitive predictive maintenance: A survey of graph-based approaches. *Manuf. Syst.* **2022**, *64*, 107–120. [CrossRef]
14. Liu, S.; Lu, Y.; Shen, X.; Bao, J. A digital thread-driven distributed collaboration mechanism between digital twin manufacturin units. *J. Manuf. Syst.* **2023**, *68*, 145–159. [CrossRef]

https://doi.org/10.3390/mi1702022

15. Zhu, Z.; Xi, X.; Xu, X.; Cai, Y. Digital Twin-driven machining process for thin-walled part manufacturing. *J. Manuf. Syst.* **2021**, *59*, 453–466. [CrossRef]
16. Plakhotnik, D.; Curutiu, A.; Zhulavskyi, A.; Beudaert, X.; Munoa, J.; Stautner, M. Framework for Coupled Digital Twins in Digital Machining. *MM Sci. J.* **2021**, *2021*, 5093–5097. [CrossRef]
17. Ma, C.; Gui, H.; Liu, J. Self learning-empowered thermal error control method of precision machine tools based on digital twin. *J. Intell. Manuf.* **2021**, *34*, 695–717. [CrossRef]
18. Liu, S.; Lu, S.; Li, J.; Sun, X.; Lu, Y.; Bao, J. Machining process-oriented monitoring method based on digital twin via augmented reality. *Int. J. Adv. Manuf. Technol.* **2021**, *113*, 3491–3508. [CrossRef]
19. Dreyfus, P.-A.; Psarommatis, F.; May, G.; Kiritsis, D. Virtual metrology as an approach for product quality estimation in Industry 4.0: A systematic review and integrative conceptual framework. *Int. J. Prod. Res.* **2021**, *60*, 742–765. [CrossRef]
20. Wu, Y.; Zhou, L.; Zheng, P.; Sun, Y.; Zhang, K. A digital twin-based multidisciplinary collaborative design approach for complex engineering product development. *Adv. Eng. Inform.* **2022**, *52*, 101635. [CrossRef]
21. Sulitka, M.; Kolar, P.; Sveda, J.; Smolik, J.A.N. Strategy for Implementating Predictive Process-Oriented Machine Tool Digital Twins. *MM Sci. J.* **2022**, *2022*, 5954–5961. [CrossRef]
22. Xin, Y.; Chen, Y.; Li, W.; Li, X.; Wu, F. Refined Simulation Method for Computer-Aided Process Planning Based on Digital Twin Technology. *Micromachines* **2022**, *13*, 620. [CrossRef]
23. Zheng, C.M.; Zhang, L.; Kang, Y.-H.; Zhan, Y.; Xu, Y. In-process identification of milling parameters based on digital twin driven intelligent algorithm. *Int. J. Adv. Manuf. Technol.* **2022**, *121*, 6021–6033. [CrossRef]
24. Chen, Z.; Zou, J.; Wang, W. Towards comprehensive digital evaluation of low-carbon machining process planning. *Artif. Intell. Eng. Des. Anal. Manuf.* **2022**, *36*, e21. [CrossRef]
25. Li, J.; Zhou, G.; Zhang, C. A twin data and knowledge-driven intelligent process planning framework of aviation parts. *Int. J. Prod. Res.* **2021**, *60*, 5217–5234. [CrossRef]
26. Liu, J.; Zhou, H.; Liu, X.; Tian, G.; Wu, M.; Cao, L.; Wang, W. Dynamic Evaluation Method of Machining Process Planning Based on Digital Twin. *IEEE Access* **2019**, *7*, 19312–19323. [CrossRef]
27. Shen, N.; Wu, Y.; Li, J.; He, T.; Lu, Y.; Xu, Y. Research on procedure optimisation for composite grinding based on Digital Twin technology. *Int. J. Prod. Res.* **2022**, *61*, 1736–1754. [CrossRef]
28. Wang, R.; Ji, D.Y.; Liu, C.; Lee, J. V-TimesNet: Vision-Augmented TimesNet for Improved Anomaly Detection in Semiconductor Plasma Dry Etching. *SSRN* **2025**, *preprint*.
29. Wu, M.; Shukla, S.; Vrancken, B.; Verbeke, M.; Karsmakers, P. Data-Driven Approach to Identify Acoustic Emission Source Motion and Positioning Effects in Laser Powder Bed Fusion with Frequency Analysis. *Procedia CIRP* **2025**, *133*, 531–536. [CrossRef]
30. Wang, R.; Cheng, M.N.; Loh, Y.M.; Wang, C.; Cheung, C.F. Ensemble learning with a genetic algorithm for surface roughness prediction in multi-jet polishing. *Expert Syst. Appl.* **2022**, *207*, 118024. [CrossRef]

https://doi.org/10.3390/mi17020226

micromachines

MDPI

Review

A Review of the Machining Mechanisms in Field-Assisted Cutting of Brittle Materials

Xuexiang Sheng [1], Zhanchen Zhu [2,*] and Changlin Liu [3,*]

1 Engineering and Technology Innovation Center, Shandong Harbor Engineering Group Co., Ltd., Rizhao 276800, China; 18063309392@163.com
2 State Key Laboratory of Ultra-Precision Machining Technology, Department of Industrial and Systems Engineering, The Hong Kong Polytechnic University, Hong Kong 999077, China
3 School of Mechanical Science and Engineering, Huazhong University of Science and Technology, Wuhan 430074, China
* Correspondence: zhanchen.zhu@connect.polyu.hk (Z.Z.); charlie_lcl@outlook.com (C.L.)

Abstract

Brittle materials such as single crystals, polycrystalline ceramics, and amorphous glass are indispensable in modern industry. Driven by improvements in equipment performance, the required fabrication precision for optical elements and devices has reached nanoscale and is steadily advancing toward atomic level. Despite their outstanding physical and chemical properties, fabricating a defect-free surface with nanometer-level roughness on brittle materials is challenging due to microcracking, brittle fracture and severe tool wear. In recent years, field-assisted cutting has emerged to overcome the bottleneck in ultra-precision cutting of brittle materials. This review summarizes investigations of material removal mechanisms of brittle materials in ultra-precision cutting and surveys representative field-assisted cutting technologies—including laser, vibration, magnetic field, and ion implantation assisted cutting—highlighting how these fields broaden ductile-regime machining and suppress the machining-induced defects. This review further discusses the emerging multi-field coupling strategies and outlines future research directions in machining mechanisms to enable high-efficiency, low-damage, and high-consistency manufacturing of brittle materials.

Keywords: ultra-precision machining; field-assisted cutting; material removal mechanism; molecular dynamics simulation

1. Introduction

Owing to the superior optical, electrical, and thermal properties, brittle materials have been widely utilized in the manufacturing of critical components in precision optics, space exploration, and biomedical engineering. Driven by the demand to improve the performance of the devices, the required manufacturing precision for brittle components has advanced to nanoscale and is steadily progressing toward atomic level. However, due to the inherent brittle nature of these materials, surface defects, subsurface damage, and rapid tool wear are often inevitable during machining. These issues have emerged as formidable bottlenecks that impede the achievement of nanometer-level surface quality of brittle materials. In recent years, researchers have demonstrated that brittle materials can be machined in ductile mode via ultra-precision machining (UPM) technologies [1]. By reducing the material removal thickness into nanoscale, the machined surface quality can be significantly improved with less generation of surface defects. Nevertheless, as the

https://doi.org/10.3390/mi17030361

material removal thickness decreases to nanoscale, the microstructure of materials and specific machining conditions, such as crystal orientation, pre-existing defects, loading rate, and machining temperature, exert a profound influence on the removal process, rendering the underlying material removal mechanisms highly complex. Furthermore, the maximum material removal thickness for ductile mode removal is quite small, making it difficult to realize large-scale industrial manufacturing.

In recent years, field-assisted cutting has become a focus of academic research in UPM. Its central idea is to introduce controllable external energy fields (such as laser heating, ultrasonic tool vibration, magnetic fields, etc.) on top of conventional cutting processes to actively regulate the local mechanical response of the material or the removal mode of the machining process. This technology can effectively broaden the conditions for ductile-regime removal of brittle materials, reduce the cutting forces, suppress surface and subsurface defects, and ultimately improve the machined surface quality and machining efficiency. The application of field-assisted cutting in UPM has achieved great success in various brittle materials. Furthermore, to amplify the effects of applied assistive fields, researchers have coupled the energy fields and explored the mechanisms in multi-field assisted cutting. This approach provides a novel technical route for enhancing the machining quality of brittle materials. In field-assisted cutting, the introduced energy field alters both the intrinsic properties of materials and the tool–workpiece interaction. Consequently, the underlying machining mechanisms differ fundamentally from those of conventional processes, making this a prominent research hotspot in the current field of UPM.

This review focuses on the research trajectory and key issues in field-assisted cutting of brittle materials, as shown in Figure 1. In Section 2, we provided an overview of representative brittle materials and their material removal mechanisms during UPM. We then discussed the development and machining mechanisms of representative field-assisted cutting technologies in Section 3. Based on a synthesis and analysis of the current state of the art, we summarized the current research in Section 4 and further discussed challenges and research directions in Section 5. Through these contents, this paper aims to provide a systematic reference for academic investigation and engineering implementation of field-assisted cutting in the high-quality manufacturing of brittle materials.

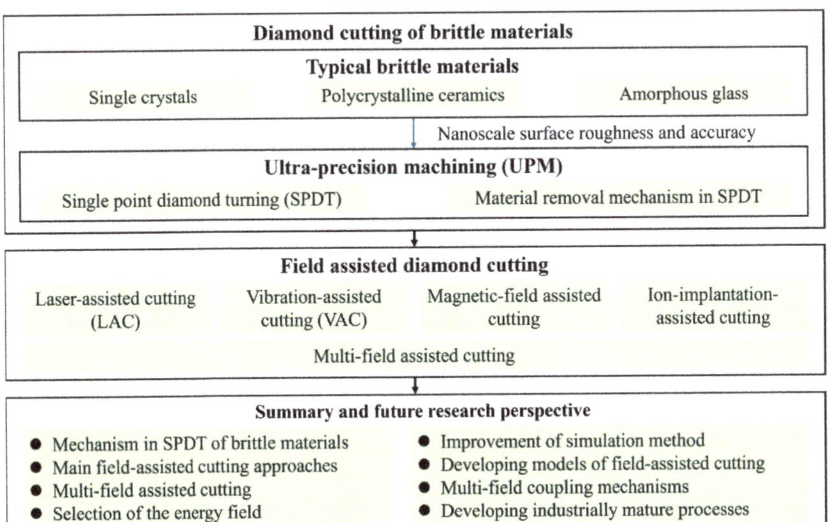

Figure 1. Structural diagram of the review.

https://doi.org/10.3390/mi17030361

2. Diamond Cutting of Brittle Materials

2.1. Typical Brittle Materials

Brittle materials refer to a class of materials that fracture with little deformation when subjected to stress (the critical strain normally being below 5% [2]). Brittle materials usually exhibit good wear resistance and corrosion resistance, high stiffness, and excellent semiconductor properties, making them key components in advancing modern technology, such as integrated circuits, precision instruments, and space optics [3–5]. Most of the inorganic non-metallic materials are brittle materials, including single crystals, polycrystalline ceramics, and amorphous glass, as illustrated in Figure 2.

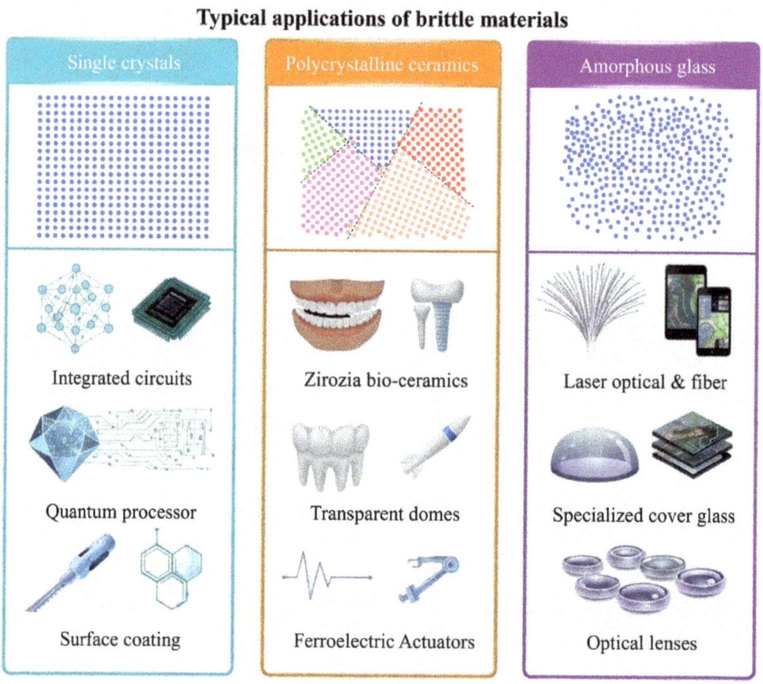

Figure 2. Typical applications of brittle materials.

(a) Single crystals

Single crystals such as silicon, silicon carbide, and quartz are materials in which atoms are linked throughout the entire structure by strong, directional covalent bonds, forming a continuous 3D (or sometimes layered) bonding network rather than discrete molecules. Because of this bonding, single crystals exhibit unique properties including high melting points, hardness and brittleness [6,7]. The arrangement of atoms in single crystals plays a crucial role in determining their stability and overall characteristics, making them distinct in the context of molecular interactions and crystal formations.

Single crystals are widely used in the fields of microelectronics, precision optics, and advanced manufacturing [8]. For example, as a highly ordered form of atoms, single-crystal silicon has uniform electrical and mechanical properties, making it the most important substrate material in modern electronics such as integrated circuits [9], power devices [10], and micro-electro-mechanical systems (MEMS) [11]. Owing to its advantages in optical and mechanical properties, single-crystal silicon is widely used for infrared windows [12],

https://doi.org/10.3390/mi1703036

X-ray mirrors [13], and components in harsh environment systems [14]. Furthermore, as a representative third-generation semiconductor, single-crystal silicon carbide is an important material in high-power and high-temperature electronics as well as optical components under harsh thermal and mechanical conditions [15]. Moreover, very hard covalent crystals such as diamond and cubic boron nitride are essential materials in abrasives [16], cutting tools [17], and protective coatings [18]. Diamond is also a prominent candidate for next-generation semiconductor devices owing to its extraordinary combination of electronic, thermal, mechanical, and quantum properties [19].

(b) Polycrystalline ceramics

Ceramic materials are a class of materials composed primarily of inorganic non-metallic compounds, typically manufactured through powder batching, forming, and high-temperature sintering. Polycrystalline ceramic is a composite whose properties reflect the properties of its grains and grain boundaries. Typical polycrystalline ceramics include alumina, zirconia, silicon nitride, and aluminum nitride. Because of the strong ionic and covalent bonding, polycrystalline ceramics generally exhibit high hardness and wear resistance, high-temperature capability and corrosion resistance, and good electrical insulation and dielectric properties.

Polycrystalline ceramics have widespread structural and mechanical applications. Owing to the high specific stiffness and dimensional stability, reaction-bonded silicon carbide is a promising material for large-aperture mirrors used in the telescope community worldwide [20]. Bio-ceramics such as alumina and zirconia are mainly used in orthopedic and dental repair due to their compatibility with the physiological environment, desired mechanical strength and wear resistance [21]. Alumina is also a key optical material used in transparent windows and domes for harsh environments such as infrared missile domes, high-pressure viewports, optical coatings, and protective sensor windows [22]. Furthermore, ferroelectric ceramics such as lead zirconate titanate, barium titanate and lead lanthanum zirconium titanate are widely used in electrocaloric cooling, electromechanical actuators, and transducers renowned for their desired properties including exceptional electrocaloric effects and high piezoelectric coefficients [23,24].

(c) Amorphous glass

Amorphous glass is an inorganic, non-metallic solid that has a non-crystalline atomic structure. It is typically produced by melting raw materials and cooling the melt fast enough to avoid crystallization, forming a rigid, strongly bonded network [25]. Glass is a good electrical insulator with relatively high chemical stability, hardness and brittleness. Unlike crystalline materials, glass has no mobile dislocations to blunt crack tips, and small surface flaws such as scratches or microcracks could create strong stress concentrations during deformation. Under tensile or bending loading, these cracks could propagate rapidly, causing sudden fracture with little or no yielding. Glass is widely used in optical systems owing to its high transparency, homogeneity, and precise controllability in refractive index and dispersion [26]. In imaging systems, optical glasses are widely used to manufacture lenses (camera, microscope, telescope, eyeglasses) and prisms that refract and shape light while minimizing aberrations [27]. In laser and photonics systems, glass serves as windows, beam splitters, and mirrors. It is also the core material for optical fibers used in telecommunications [28] and hemispherical resonator gyroscopes in navigation systems [29]. Specialized glasses are additionally used for UV optics, infrared-transmitting optics, and protective cover plates for displays and detectors where optical clarity and surface quality are critical [30].

Driven by the demand for improving device performance in these fields, precision components are facing increasingly stringent requirements for machining quality and material properties [31,32]. High-end equipment often demands nanometer-level surface

https://doi.org/10.3390/mi17030361

roughness and extremely low subsurface machining damage [33,34]. However, brittle materials face significant difficulty in fabricating high-quality surfaces with nanoscale surface roughness and low subsurface damage. During the machining process, cracks tend to initiate and surface defects tend to propagate rapidly, leading to sudden fracture with little to no deformation in the workpiece material [35,36]. Furthermore, rapid cutting tool wear, fluctuations in machining forces, pronounced thermal effects, and significant residual stresses further raise the threshold for achieving nanometer-level surface quality. Therefore, achieving high-precision, low-damage machining of these brittle materials has become a major research focus in the field of advanced manufacturing.

2.2. Ultra-Precision Cutting Technology

To meet the machining quality requirements of components used in high-end equipment, UPM technologies—including ultra-precision scratching [37,38], cutting [39], grinding [40,41], and polishing [42,43]—have gradually become indispensable methods in advanced manufacturing. Among these technologies, ultra-precision cutting represented by single-point diamond turning (SPDT) has become the preferred technique for UPM of complex structures owing to its advantages of high flexibility, high accuracy, and high machining efficiency [44]. During ultra-precision cutting, the workpiece is fixed on a high-precision air/hydrostatic spindle and rotated at a controlled speed [45]. A diamond tool is mounted on ultra-precise linear slides and moved under CNC control. During machining, the spindle rotates to provide the primary cutting motion while the tool follows a programmed tool path to generate feed and the desired surface geometry. Relying on nanometer-scale positioning, excellent spindle error motion control, high machine stiffness, and a sharp cutting edge, optical-quality surfaces with high form accuracy and low surface roughness can be achieved. In its early stage of development, ultra-precision cutting was mainly applied to machine ductile metals such as aluminum and copper. With the advancement of space optics and defense systems, the demand has grown rapidly for ultra-precision optical components made from brittle materials such as single-crystal silicon and silicon carbide. This demand has driven the expansion of ultra-precision cutting into machining of brittle materials.

2.3. Removal Mechanisms of Brittle Materials

Unlike ductile metals, brittle materials have high hardness and low plasticity. Surface defects such as micro-chipping and cracking are generally inevitable during the machining process [46,47]. Conventional processes therefore struggle to produce surfaces with high accuracy and low subsurface damage. Based on extensive cutting experiments, scholars have revealed that there are three typical characteristics of material removal when brittle materials are cut by decreasing the removal thickness: brittle removal, brittle–ductile transition, and ductile removal, which can be clearly identified by plunge-cutting experiments, as illustrated in Figure 3a. In the brittle removal stage, materials are mainly removed by cracks, chipping, and fractures with irregular plastic deformation, generating surfaces with severe defects and subsurface damage, as shown in Figure 3c. As the removal thickness decreases, the material removal mechanism could gradually transit into ductile mode, making it possible to fabricate high-quality surfaces of brittle materials, as shown in Figure 3d,e. In 1989, Puttick et al. [48] claimed the existence of a critical material removal thickness during SPDT below which the brittle material can be removed by stable plastic deformation, leaving a crack-free machined surface. In the 1990s, Blake and Scattergood [49] introduced the concept of critical material removal thickness for hard and brittle materials through cutting experiments. They proposed a feed model to determine the geometric relationship between the tool and the workpiece and the ductile–brittle transition in SPDT, as shown in Figure 4.

https://doi.org/10.3390/mi17030361

Based on this model, the ductile–brittle transition is closely related to tool geometry and cutting parameters, which is further confirmed by Cai et al. [50]. Yan et al. [51] pointed out that the high compressive stress introduced by decreasing the tool rake angle or material removal thickness is an important reason for ductile machining of hard and brittle materials. Afterwards, the ductile machinability of various brittle materials has been confirmed in SPDT [52].

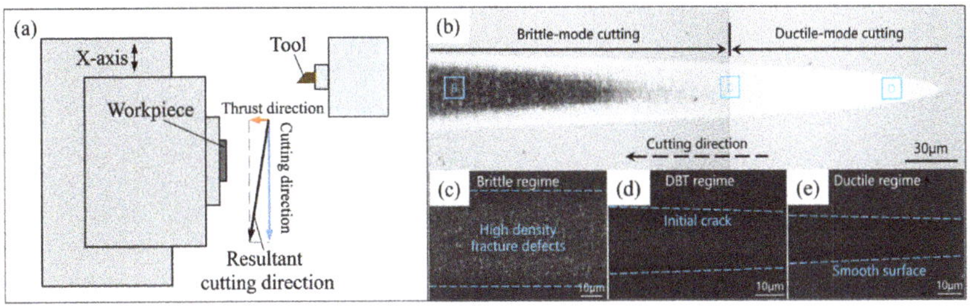

Figure 3. Determination of the removal characteristics by plunge-cutting experiments. (**a**) Illustration of the plunge-cutting experiments. (**b**) Surface morphology of the machined surface from the plunge-cutting experiment, and (**c–e**) micrograph of the surface from brittle, transition, and ductile mode machining [53].

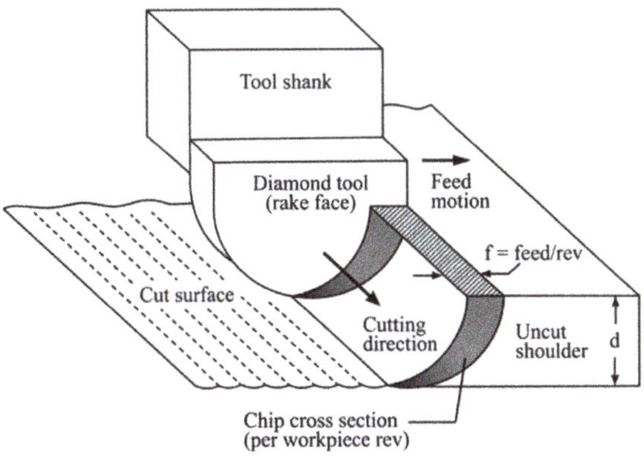

Figure 4. Geometric feed model for ultra-precision cutting.

In macroscale cutting of metallic materials, the material removal thickness is typically on the micrometer scale, which is much larger than the tool edge radius. Therefore, the material removal process is mostly described using the classical shear model without considering the size effect of the cutting tool edge [54]. In the ductile removal stage of ultra-precision cutting, the material removal thickness decreases to the nanoscale, which is in the same order of magnitude as the tool edge radius. Therefore, the size effect of the cutting tool edge has a pronounced influence on the material removal behavior. Due to the difficulty in direct observation of material deformation at the nanoscale, numerical simulation has become a preferred method to explore the mechanisms during UPM. Scholars have conducted extensive numerical investigations to reveal the material removal behavior, subsurface structure evolution, and cutting tool wear in UPM via finite element (FE) simulation [55], discrete element method (DEM) [56], smoothed particle hydrodynamics (SPH) [57], and molecular dynamics (MD) simulation [58]. In particular, MD simulation, which is based on first-principles calculations or empirical interatomic potentials, has been widely used to describe the material deformation behavior in UPM owing to its advantages

https://doi.org/10.3390/mi17030361

in simulating plastic deformation behavior such as dislocation propagation and phase transition [59–61]. In the 1980s, Lawrence Livermore National Laboratory first applied MD simulation to investigate the material deformation mechanism of single-crystal copper. In the early stages, such studies were limited in the size of the simulation system due to computational constraints. Nowadays, with the development of computer technology and interatomic potentials, MD simulation has gradually become a primary approach for studying the nanoscale cutting mechanisms of brittle materials including single crystals [62], polycrystalline ceramics [63], and amorphous glass [64].

For single crystals, phase transition and dislocation slip are two dominant mechanisms of plastic deformation. During ultra-precision cutting, the high compressive stress introduced by the tool edge causes localized structure evolution in the tool–workpiece contact region. Fang et al. [65] first proposed that the dominant material removal mechanism in ultra-precision cutting of single-crystal silicon is extrusion, as shown in Figure 5. They suggested that the compressive stress near the tool edge causes high-pressure phase transformation (HPPT) of the cubic diamond phase into the high-pressure phase (e.g., the Si-II phase) in the deformation zone [66], thereby enabling the extrusion of the deformed material during cutting [67]. Another feature of extrusion is the existence of a so-called stagnation region ahead of the tool edge [68], where the workpiece atoms are nearly stationary relative to the cutting tool during the cutting process. Atoms above the stagnation region are piled upward to form chips and ridges through extrusion, while atoms below the stagnation region are compressed downward into the workpiece. After cutting, the compressed material undergoes further transformation into an amorphous phase, forming an amorphous layer that remains on the machined surface [69].

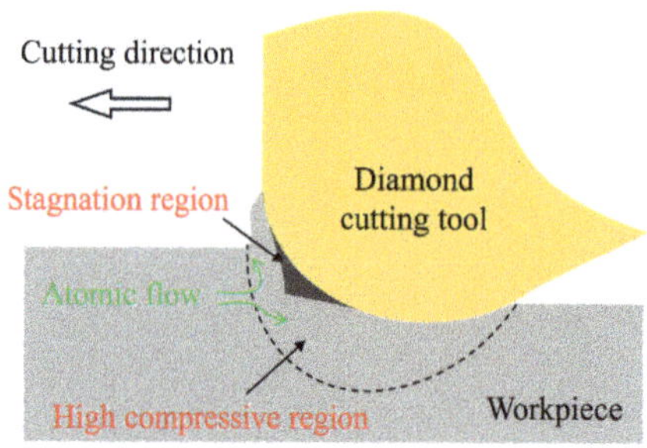

Figure 5. Schematic illustration of the material removal behavior during ultra-precision cutting.

When the material removal thickness increases, the removal mechanism of single crystals shows new features other than extrusion. Kausala and Zhang [70] found that for single-crystal silicon, nanotwins can be observed in the subsurface workpiece when the scratch depth exceeds 1 nm, which indicates the existence of slip motion during cutting. When the material removal thickness further increases, the deformation mechanism shows some features similar to those of macroscale cutting. Based on an MD cutting simulation with a removal thickness increasing to 20 nm, Liu et al. [71] suggested that the material removal mechanism shifts from extrusion to shearing as the cutting tool edge radius decreases. This transition occurs when the tool edge radius is approximately equal to the material removal thickness. To quantitatively characterize this relationship, Yan et al. [72] proposed the relative tool sharpness (RTS), which is defined as the ratio between tool edge radius and material removal thickness. They pointed out that as RTS increases, the dominant removal mechanism of single-crystal silicon transitions from rubbing ($0 < RTS < 0.1$) to ploughing ($0.1 < RTS < 0.25$), then to extrusion ($0.25 < RTS < 0.5$), and finally to shearing ($0.5 < RTS$) [72]. This transition was further observed for other single crystals such as silicon carbide [73]. In addition to tool edge radius, Liu et al. [74] pointed out that the tool rake angle and crystal orientation of workpieces have a significant influence on the dominant material removal mechanism of single crystals, as illustrated in Figure 6.

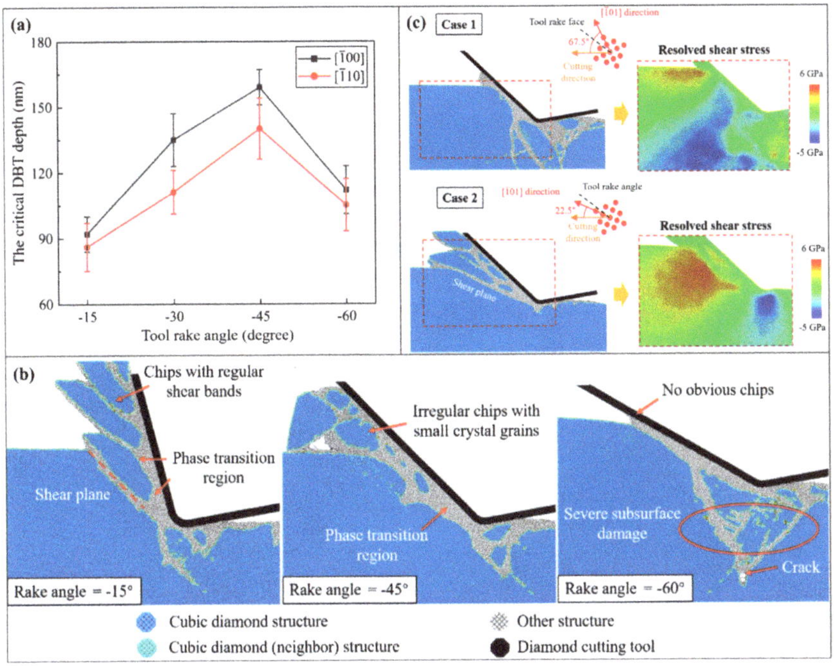

Figure 6. The material removal behavior of single-crystal silicon. (**a**) Influence of tool rake angle on the critical depth of cut for ductile removal. (**b**,**c**) Material removal behavior under different tool rake angles and crystal orientations of workpieces [74].

Unlike single crystals, polycrystalline ceramics present complex material removal mechanisms due to the randomly distributed crystal grains and grain/phase boundaries [75]. Multiple material deformation and removal mechanisms coexist and interact during the cutting process, and their dominance depends strongly on the machining conditions and microstructural characteristics. The random arrangement of grains and boundaries indicates that polycrystalline ceramics can be effectively modeled as an assemblage of bonded discrete particles, making numerical approaches such as DEM and MD particularly suitable for studying the deformation features during machining [76,77]. At small material removal thickness, plastic deformation can occur locally, and the material removal mechanism is similar to that of single crystals [78]. Nevertheless, because of the discrete bonding between grains, friction between the tool and the workpiece could introduce micro-cracks near the grain/phase boundaries [79]. Meanwhile, breakage of intergranular bonds leads to intergranular fracture and grain boundary separation [80], making it difficult to fabricate a defect-free surface on polycrystalline ceramics even when the single-crystal components are removed in ductile mode [81]. In practical machining, polycrystalline ceramics usually exhibit mixed ductile–brittle material removal features. Smooth regions produced by plastic deformation or amorphization coexist with fractured regions caused by intergranular cracking [82], as shown in Figure 7. With the increase in material removal thickness, crack propagation could appear inside the grains, causing brittle mode removal that is similar to that of single crystals, coupled with intergranular fracture and grain boundary separation.

https://doi.org/10.3390/mi17030361

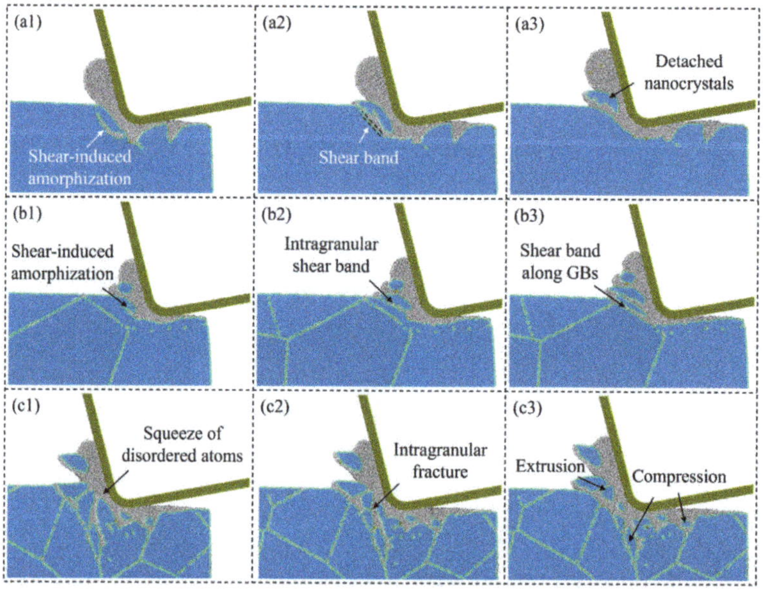

Figure 7. Snapshot of the material removal behavior during cutting on: (**a1–a3**) a single-crystal workpiece; (**b1–b3**,**c1–c3**) polycrystal workpieces with different grain sizes [82].

For amorphous glass, the machining mechanism is distinct from that in crystalline materials, since atoms in glass are not arranged in periodic lattices, which fundamentally limits conventional plastic deformation mechanisms such as dislocation motion and phase transition. The three-dimensional network formed by randomly arranged and interconnected atoms contains massive atomic-scale voids and dangling bonds, resulting in a relatively low packing density of atoms [83]. During UPM, amorphous glass undergoes localized plastic deformation through densification and shear flow of the glass network at small material removal thickness [84], which contributes to ductile-mode material removal. Taking fused silica as an example, bonds between silicon and oxygen atoms undergo twisting and bending to fill the voids during machining, leading to variations in nanoscale structure such as n-membered rings [64] and densification [85]. Furthermore, fused silica exhibits localized plastic flow due to surface friction and shear stress under compression and shear deformation (shown in Figure 8), which contributes to pile-up and material removal during the machining process [86,87]. When the material removal thickness increases, median and radial cracks may initiate due to stress concentration, which is an important reason for the transition to micro-brittle fracture and chipping [88].

In summary, the ultra-precision cutting technologies represented by SPDT have enabled ductile machining to fabricate defect-free surfaces of a variety of brittle materials. Scholars have conducted extensive in-depth investigations into the underlying mechanisms of the machining process through experiments and numerical approaches such as MD. It is worth mentioning that due to limitations in the accuracy of interatomic potential and computational capacity, discrepancies still exist between MD simulations and experiments in terms of simulation scale and machining parameters. Furthermore, tool wear and subsurface damage caused by the inherent hardness and brittleness of materials, as well as the relatively low ductile removal thickness that results in limited machining efficiency, remain critical bottlenecks restricting improvements in quality and efficiency of UPM for optical components. As the research into the nanometric cutting mechanism of brittle materials continues to deepen, enhancing ductile machinability, suppressing cutting tool wear, and

https://doi.org/10.3390/mi17030361

reducing subsurface damage have become the key issues in improving the manufacturing capability of brittle optical components.

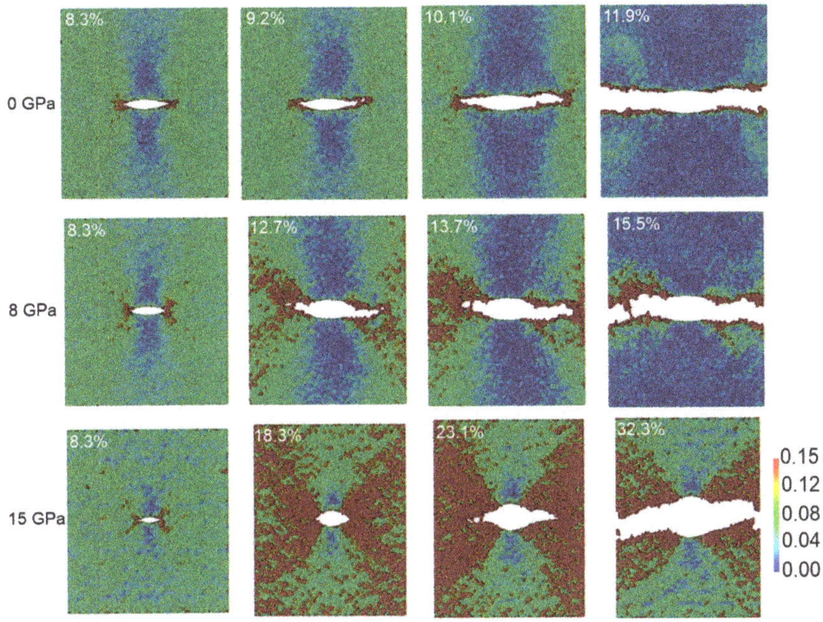

Figure 8. Local shear strain mapping of a V-crack tension test under different stresses of fused silica glass via MD simulation [89].

3. Field-Assisted Cutting Technology

At present, it is difficult to further improve the machining quality solely by adjusting the machine tools and process parameters. In recent years, field-assisted cutting technologies, such as laser heating, tool vibration, and magnetic field, have provided new approaches to enhance the machinability of brittle materials and attracted increasing attention from the academic community. Figure 9 illustrates the development of typical field-assisted cutting technologies [90–92]. This section introduces the development and mechanisms of various field-assisted cutting technologies commonly used in UPM of brittle materials.

3.1. Laser-Assisted Cutting

Since the hardness of materials can be decreased by raising the temperature, thermal field can be used as an auxiliary approach to improve the machinability of hard and brittle materials [93]. In the 1950s, thermal assistance had already emerged in the machining of hard-to-machine materials to achieve better machining performance. It has been proven that with thermal assistance, higher material removal rate [94], better surface quality [95] and lower cutting tool wear [96] can be achieved for hard and brittle materials, as shown in Table 1. During the machining process, a highly controllable heat source is crucial to control the thermally influenced region. Common heat sources in practical applications include plasma [97], laser beams [98], gas heating [99], and induction heating [100]. With the advancement of laser technology, laser-assisted cutting (LAC) has become a research hotspot in the field of thermal-assisted machining owing to its advantages of high positioning accuracy and energy density of the thermal field. In the 21st century, in situ laser-assisted diamond turning was proposed by Shayan et al. [101]. By adjusting a laser beam that

https://doi.org/10.3390/mi17030361

passes across the diamond tool edge, precise control of the laser energy can be achieved, as illustrated in Figure 10b. This heating method allows precise control of the heat distribution within the workpiece, avoiding the overheating caused by conventional laser heating (shown in Figure 10a) and thereby effectively suppressing thermal expansion and thermal damage of the workpiece [102]. At present, this technique has been successfully applied to the ultra-precision cutting of various brittle materials such as single-crystal silicon [103], fused silica [104], silicon carbide [105], sapphire [106], and silicon nitride ceramic [107].

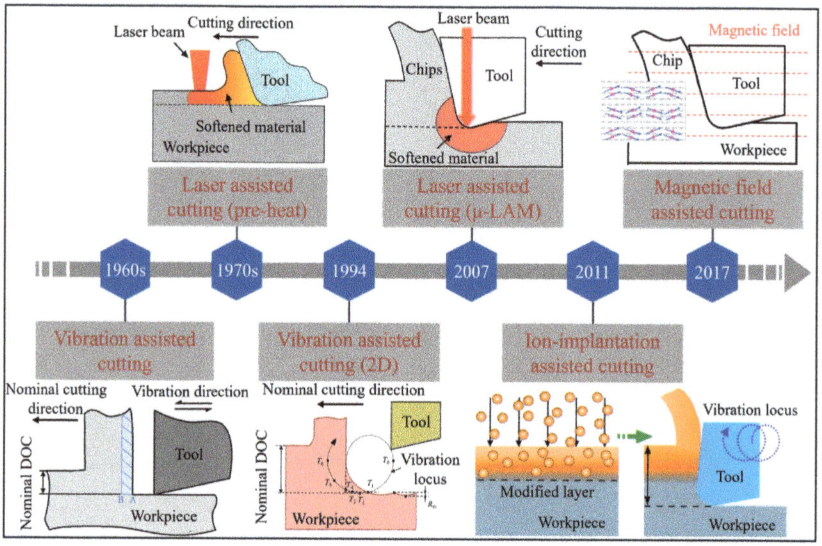

Figure 9. Development of typical field-assisted cutting technologies.

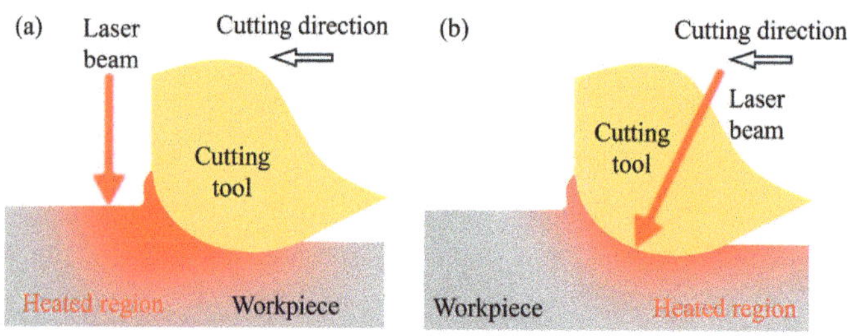

Figure 10. Schematics of the heating strategy in LAC. (**a**) Pre-heat LAC. (**b**) In situ LAC.

Table 1. Comparison between conventional cutting and laser-assisted cutting of several materials (conventional/laser-assisted cutting).

Material	Machined Surface Roughness (nm)	Critical Depth of Cut (nm)
Silicon	Sa: 6.006/2.630 [103]	150/395 [108]
Germanium	Ra: 12.756/9.898 [109]	42.095/211.286 [109]
Magnesium fluoride	Sa:21.289/3.911 [110]	555.23/1079.30 [110]
Fused silica	Sa: 113.362.714 [111]	35.519/112.638 [111]

During LAC, the machining mechanism differs from that of conventional cutting since the material properties and deformation features are significantly affected by temperature [110,112]. For brittle materials, the cutting forces and internal stress in the deformation zone [113] decrease significantly with increasing temperature due to the thermal

https://doi.org/10.3390/mi1703036

softening effect, contributing to better machining performance and lower abrasive tool wear [114]. By plunge-cutting experiments with an in-situ laser heating module shown in Figure 11a, Chen et al. [108] confirmed the improvement of the machinability of single-crystal silicon by LAC, as shown in Figure 11b–e. Their results indicate that the critical depth of cut for single-crystal silicon can be improved from 150 nm to 395 nm with the assistance of laser heating. Furthermore, the phase transition behavior and dislocation motion change markedly under high-temperature conditions, causing material removal behavior that is distinct from that of conventional cutting. Because the material is removed at the nanoscale with high strain rate and temperature, MD simulation is a suitable tool for exploring the removal mechanism in LAC. In MD simulation, the laser heating process can be simulated either by a specific heating region [115,116] or heating the whole workpiece to a desired temperature obtained by FE simulation [91], as shown in Figure 12. Based on previous research, atomic flow in the deformed region is promoted when the temperature rises, which dissipates more strain energy and suppresses the elastic recovery on the machined surface [91]. For single-crystal silicon, the extrusion removal is suppressed since the internal compressive stress is decreased and fewer amorphous atoms are generated [115]. Furthermore, Chen et al. [108] found that the amorphous phase can be generated directly from the cubic diamond phase via lattice collapse without the formation of intermediate high-pressure phases. The newly formed amorphous phase partially crystallizes during cutting and transforms into the Si-III and Si-XII phases under annealing. In addition to phase transition, they suggested that laser heating promotes dislocation activity, thereby greatly enhancing cleavage and shear removal at high temperatures [117,118]. Similar results were reported by Meng et al. [119], and they pointed out that increasing the machining temperature is beneficial for triggering the nucleation and extension of dislocations in the glide slip system, which is sensitive to high temperature.

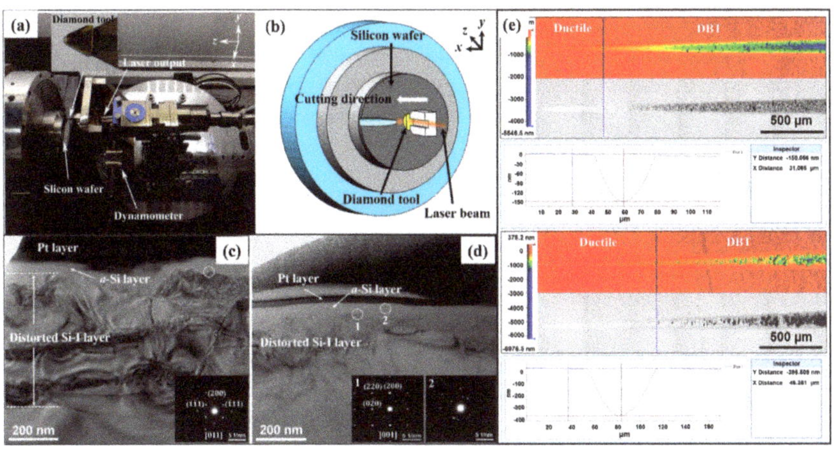

Figure 11. LAC for single-crystal Si. (**a**,**b**) Experimental setup of LAC. (**c**,**d**) TEM images of the subsurface workpiece in ductile cutting regions. (**e**) Improvement of the critical depth of cut by applying LAC [108].

For crystalline materials, laser heating introduces crystallization of the disordered atoms beneath the machined surface, which is beneficial for reducing the subsurface damage generated during machining. As early as 1979, researchers found that the crystallization of amorphous silicon is closely related to temperature [120]. During ultra-precision cutting, the disordered phases could recrystallize into a crystalline structure with increasing temperature [121]. For silicon, this process can be promoted by increasing compressive

https://doi.org/10.3390/mi17030361

stress during subsequent processing, since the cubic diamond phase has a close-packed atomic structure and higher density than the amorphous phase [122]. However, recrystallization into other phases is usually inevitable during LAC. Langana et al. [123] found that excessive heating power introduces multiple crystalline phases on the machined surface, accompanied by undesired residual stress. From MD simulation, Chen et al. [124] reported the existence of recrystallization into a hexagonal structure of single-crystal silicon during annealing of the machined workpiece at high temperature. For polycrystalline ceramics, different crystallization features can be observed due to the existence of grain boundaries and mismatch in material properties between phases. Liu et al. [125] revealed that the recrystallization of silicon is more apparent than that of silicon carbide during LAC of reaction-bonded silicon carbide.

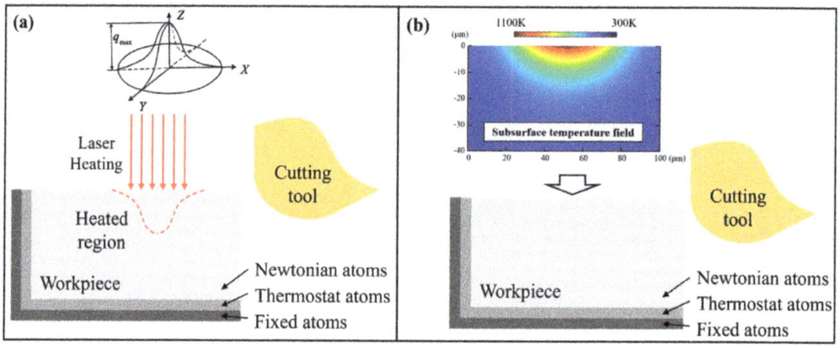

Figure 12. MD model for LAC where (**a**) the laser heating is simulated by adding a specific heating region and (**b**) the entire workpiece is heated to the desired temperature according to FE simulation of the temperature distribution [91].

The above-mentioned studies have demonstrated the advantages of LAC in machining brittle materials, as well as the laser-induced changes in material removal behavior and subsurface damage mechanisms. However, how laser heating influences the subsurface structure is mainly investigated by MD simulation and post-process observation, which lacks convincing evidence of the dynamic evolution of microstructure during machining. Furthermore, regardless of the benefit in tuning the material properties of the workpiece, laser heating could reduce the hardness and increase the chemical reactivity of the cutting tool [126], which may lead to severe chemical tool wear [127]. Therefore, more detailed investigation of the thermal softening effect should be conducted, and LAC must strike a balance between workpiece softening and tool softening to achieve optimal machining quality.

3.2. Vibration-Assisted Cutting

Vibration-assisted cutting (VAC) has been applied in the manufacturing industry since the 1960s. In its early applications, the cutting tool undergoes one-dimensional linear vibration along the nominal cutting direction, which is also known as linear vibration cutting (LVC) [128], as illustrated in Figure 13a. Compared with conventional cutting, LVC effectively reduces the cutting forces [129], mitigates process instability [130], and extends cutting tool life [131]. However, the tool's linear vibration tends to imprint vibration marks on the machined surface, thereby degrading the surface quality. To address this issue, Shamoto and Moriwaki proposed two-dimensional elliptical vibration cutting (EVC) [132], in which the cutting tool vibrates in two dimensions within the plane defined by the nominal cutting direction and the nominal depth of cut direction, as illustrated in Figure 13b. During EVC, the instantaneous material removal thickness is much smaller than the

https://doi.org/10.3390/mi17030361

nominal depth of cut, and the cutting force acting on the cutting tool is markedly lower than that in conventional cutting. Furthermore, owing to the periodic contact and separation between the tool and the workpiece, most of the heat generated during cutting is removed by air and cutting fluid. At present, EVC has been successfully applied in UPM of a wide range of brittle materials, including single-crystal silicon [133], sintered silicon carbide [134], tungsten carbide [135], calcium fluoride [136], and composite materials [137].

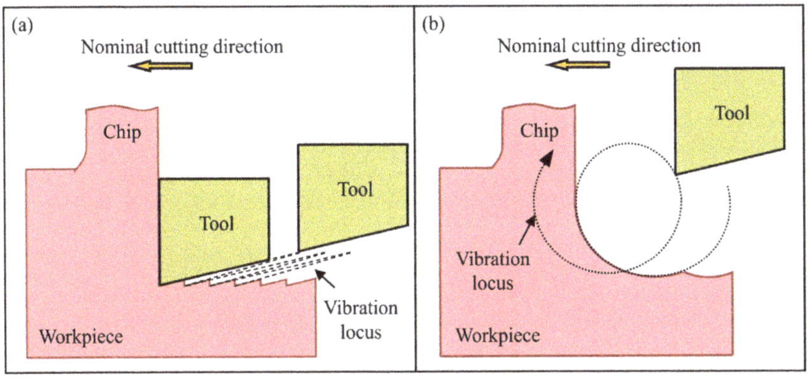

Figure 13. Schematic illustration of (**a**) LVC and (**b**) EVC.

During EVC, the high-frequency tool vibration introduces motion along the depth of the cut direction, rendering the machining mechanism more complex than that in conventional cutting. Scholars have investigated the mechanisms of surface formation and material removal of various brittle materials in EVC and demonstrated the significance of tool vibration on cutting performance [138,139]. It is revealed that the tool vibration-introduced reduction in material load is responsible for the decrease in cutting forces [140] and internal stress [141]. Furthermore, the variation in transient cutting tool motion has a significant influence on the machining anisotropy when cutting single crystals. Chen et al. [142] suggested that the tool vibration dynamically redistributes and rebalances the existing anisotropy among the applied forces, thereby leading to unique machining characteristics, including homogeneous deformation. Moreover, the tool piling-up motion facilitates the upward motion of materials, which tends to enhance the atomic flow and increase the material removal rate [143]. The tool vibration-introduced upward motion causes an undesired increase in tensile stress during the cutting process, which causes tearing of brittle materials such as single-crystal silicon [144]. Therefore, investigation into the mechanisms and optimization of vibration parameters such as nominal cutting speed, vibration amplitude, and vibration frequency constitutes an important research direction in VAC.

The nominal cutting speed is considered one of the most essential parameters in EVC [145]. It is revealed that the decrease in nominal cutting speed causes an obvious reduction in material load, which is beneficial for suppressing tool wear and extending tool life [146]. Meanwhile, nominal cutting speed is an important parameter in determining the surface quality since it determines the theoretical surface roughness [147]. Nath and Rahman [148] found that a lower nominal cutting speed can achieve a better surface finish, which is further confirmed by Huang et al. [149]. In addition to the nominal cutting speed, the vibration amplitude and frequency also have a significant influence on the machining performance. From the results of Dai et al. [150], a lower vibration frequency and a smaller amplitude ratio can increase the material removal rate when machining single-crystal silicon. Liu et al. [151] pointed out that the vibration amplitude affects the deformed region by influencing the piling-up motion of the workpiece material. Their simulation results indicate the existence of an optimal ratio of vibration amplitude in cutting single-crystal silicon, which has been verified by experiments [144]. For the material removal process, it is suggested that the shear angle increases and the chip thickness decreases as the frequency or amplitude increases [152], which promotes shear removal of workpiece material. Furthermore, Li et al. [153] suggested that the increase in normal amplitude enhances discontinuous processing while the increase in normal vibration amplitude promotes the elastic recovery on the machined surface, which decreases the material removal efficiency.

At present, MD simulation is the dominant method for the explanation of the transient material removal mechanism in EVC. Due to limitations in model size, MD simulations of UPM typically employ models with material removal thickness ranging from several to tens of nanometers. Because the tool vibration in EVC leads to an instantaneous material removal thickness much smaller than the nominal depth of cut, prior MD models for EVC simulation commonly assume an instantaneous material removal thickness of only 1–2 nm [151], which impedes the accurate description of the material removal mechanism. To overcome this problem, Liu et al. [154] proposed a cutting model to simulate the contact stage in one vibration cycle, which effectively enlarged the transient material removal thickness to ~10 nm with a much lower instantaneous tool velocity, as shown in Figure 14a. They revealed that the transient deformation mechanism of single-crystal silicon experiences the transition from extrusion to shearing in one vibration cycle by analyzing the transient material removal process shown in Figure 14b. Based on this model, they investigated the extension of subsurface damage and discussed the variation in resolved shear stress and dislocation propagation of single-crystal silicon during EVC [90]. A main drawback of this model is that it failed to evaluate the machined surface morphology since only one vibration cycle is involved in the simulation. Therefore, establishing a more realistic simulation system is important to fully reveal the surface formation mechanism during EVC.

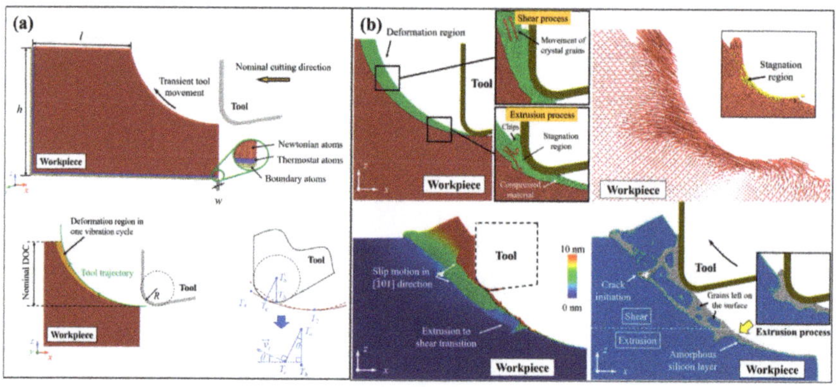

Figure 14. EVC model for simulating the contact stage in one vibration cycle [154]. (**a**) Illustration of the cutting model. (**b**) Transition of the material removal mechanism.

3.3. *Magnetic Field-Assisted Cutting*

In addition to thermal and vibration assistance, magnetic field-assisted cutting is another field-assisted cutting method, with industrial applications dating back as early as the 1960s. Compared to other energy fields used in UPM, magnetic fields offer several advantages, including low cost, simple operation, and ease of removal. In 2017, Yip and To proposed the first magnetic field-assisted cutting technology [155], aiming to minimize the material swelling effect during ultra-precision cutting of titanium alloys. Their results indicated that the magnetic field effectively increases the material removal rate and machined surface quality in the fabrication of smooth surfaces and micro-structured surfaces [156], as shown in Figure 15. This technique is mainly applied in machining magnetically susceptible materials, and it has achieved good results in the machining of brittle materials such as calcium fluoride [157]. It is revealed that a weak magnetic field introduces magneto-plasticity via the formation of non-singlet electronic states in defected configurations for brittle materials, which indicates that the magnetic field could affect the machining performance of non-magnetic brittle materials [158]. Guo et al. [159] suggested that the application of a magnetic field reduces the internal stress and facilitates the dislocation activity during machining, which is responsible for the enhancement in magneto-plasticity. In addition to the material property, the eddy-current damping effect under a magnetic field can effectively suppress vibrations in the machining system and improve the stability during the cutting process [160], which is advantageous for improving the machined surface quality. Although magnetic field-assisted cutting offers numerous advantages, its current application is restricted to a narrow range of materials due to the material-selective nature of magnetic fields. Furthermore, magnetization can adversely affect the service performance of components, which deteriorates the performance of devices in fields such as aerospace and precision metrology. Therefore, magnetic field-assisted cutting is generally not used, or demagnetization measures are employed after machining to ensure the functional performance of the machined components [161].

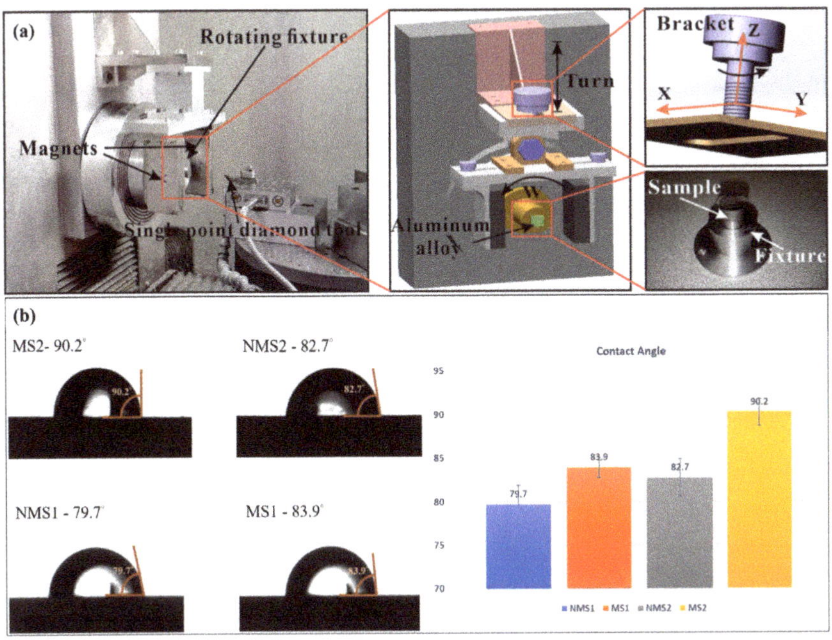

Figure 15. Magnetic field-assisted diamond turning. (**a**) The experiment setup. (**b**) The contact angle measurements of machined micro-structured surfaces [156].

3.4. Ion Implantation-Assisted Cutting

Ion implantation is a surface-modification technique that alters the physicochemical and biological properties of materials by accelerating high-energy ions and implanting them into the surface. Compared with other doping strategies, ion implantation can introduce almost any element into the target material without additional impurity elements. During

https://doi.org/10.3390/mi17030361

the implantation process, the accelerated high-energy ions are implanted into the target material. After multiple collisions, the energy is transferred to target atoms or electrons, causing structural damage that is closely related to the implantation parameters such as ion species, implantation energy, and dose. Owing to the collision-cascade effects, one implanted ion can create tens of thousands of point defects such as vacancies and interstitials in the target material. With increasing implantation dose, defect clusters such as amorphous phases can be generated in materials [162]. These defects can be used to tune the properties of the target material [163] to improve the machinability of brittle materials.

The damage structures introduced by ion implantation reduce the hardness and the brittleness of material, making it possible to mitigate brittle fractures and promote plastic deformation during UPM. In 2009, Xu et al. [164] developed a fabrication method combining focused ion beam implantation with gas-assisted etching. They successfully produced a series of micro/nanostructures such as nanogratings, nanoelectrodes, and sinusoidal microstructures. In 2011, ion implantation-assisted cutting was proposed by Fang et al. [165]. In this method, fluorine ions were implanted into single-crystal silicon to form an amorphous layer on the surface, after which ductile-regime material removal could be enabled with the ductile–brittle transition depth significantly increased to 923.566 nm. To reveal the interaction mechanism between high-energy ions and workpiece material, extensive experimental and MD investigations have been conducted owing to their advantages in simulating the structural evolution during ion implantation, as illustrated in Figure 16. Based on three-dimensional MD simulations, Chen et al. [166] proposed that the implantation causes lattice transformation into amorphous phase of single-crystal silicon, which is beneficial for reducing brittleness and fractures during machining. Furthermore, the generated amorphous phase could absorb shear strain energy and make the material more prone to plastic deformation, thereby increasing the critical depth of cut [167]. To obtain the ion distribution after implantation, To et al. [168] established a distribution model of implanted hydrogen ions and visualized the displacement damage induced in the subsurface of single-crystal silicon. These studies validated the effectiveness of ion implantation surface modification in improving the machinability of single-crystal silicon.

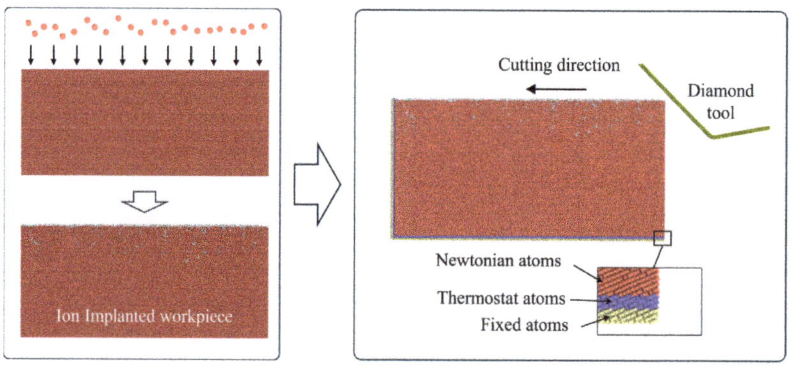

Figure 16. MD simulation of ion implantation-assisted cutting.

In recent years, researchers have further broadened the application scope of ion implantation-assisted cutting. For example, Fan et al. [169] studied the effects of implantation doses on the cutting performance of silicon carbide. Their results indicated that ion implantation significantly improved the machinability of silicon carbide, which is similar to the results from Liu et al. [170] and Tanaka et al. [171]. They also reported that even in the brittle mode, microcrack propagation is suppressed at the interface between the amorphous layer and the crystalline substrate after ion implantation. Using an in-situ scanning electron

https://doi.org/10.3390/mi17030361

microscopy observation setup, Xu et al. [172] confirmed the simulation results and found that implantation-induced damage helps to enhance the material's plastic-forming capability during the machining of brittle materials. To enhance the performance of implantation, Fan et al. [173] proposed a dual-ion implantation method that, after high-temperature annealing, can achieve a higher color-center yield within a depth of 5 nm than that obtained with helium-ion implantation alone.

In practical machining, the thickness of modified layers needs to be greater than the height of the structures to be fabricated. However, with a single ion-beam irradiation process, achieving a thick modified layer requires extremely high ion doses and long implantation times. To maximize the thickness of the modified layer, Wang et al. [174] proposed a technique that combines multiple implantations with diamond cutting in 2019, introducing sufficient lattice damage to improve the machinability of single-crystal Si. This method can significantly reduce the required ion dose, offers substantial flexibility in process-parameter design, and effectively improves machining efficiency. Their results showed that the ductile–brittle transition depth of single-crystal silicon increased from 39.7 nm to 472 nm, and high-quality micropillar arrays were successfully fabricated in the ductile mode via fly cutting. Furthermore, Tan et al. [175] applied this idea to single-crystal germanium and increased the maximum implantation depth of copper ions to 1.6 μm. After implantation, the ductile–brittle transition depth of germanium exceeded 200 nm, and the modified layer effectively suppressed subsurface damage during cutting of the ion-implanted germanium samples, ultimately enabling the successful fabrication of crack-free microlens arrays.

3.5. Multi-Field-Assisted Cutting

Although the machining performance of brittle materials can be significantly improved by field-assisted cutting, a single field has drawbacks in treating difficult-to-cut materials with specific properties such as extremely high hardness or multiple phases. In recent years, multi-field-assisted cutting technologies have been developed to address the bottlenecks of single fields in UPM [176]. Wang et al. [177] conducted a fundamental study on the combined effects of ion implantation and EVC in the cutting of binderless tungsten carbide. They found that elliptical vibration negatively affects the machinability of superhard materials, whereas ion implantation induces amorphization in the near-surface layer, reducing the strength and brittleness of tungsten carbide. This promotes chip formation via shear deformation rather than brittle fracture and effectively suppresses tool wear. Ke et al. [92] further used this method in ultra-precision cutting of sapphire. They used high-energy ions to modify the workpiece surface before cutting, as shown in Figure 17. Their results indicate a great enhancement in the machinability of sapphire by combining ion implantation with tool vibration. While Zhang et al. [178] coupled LAC and EVC to fabricate silicon microlens arrays. Their results suggest that the combination of the single assistive fields shows great advantages for the fast fabrication of microstructure arrays on brittle materials.

In multi-field-assisted cutting, the material deformation behavior is more complex than that in conventional cutting and single-field-assisted cutting, as the coupling effect of the assistive fields involves sophisticated interactions between workpieces and tools. For example, Liu et al. [179] investigated the coupled effect of LAC and EVC on the material removal mechanism of single-crystal silicon. Their results indicate that the thermal softening effect and the variation in material deformation introduced by tool vibration result in a significant improvement in the machinability of brittle materials. However, the upward motion driven by tool vibration causes undesirable rolling of detached grains, which is more apparent under high-temperature conditions. These findings demonstrated

https://doi.org/10.3390/mi17030361

the feasibility of multi-field assistance in improving the machinability of brittle materials and the coupled effect of multi fields on material removal behavior. However, the material properties and deformation behavior under multiple fields are complex, and optimization of the machining process requires the coupling of more field parameters. Therefore, further simulation and experimental investigations are needed to fully reveal the machining mechanism and broaden the application of multi-field-assisted cutting.

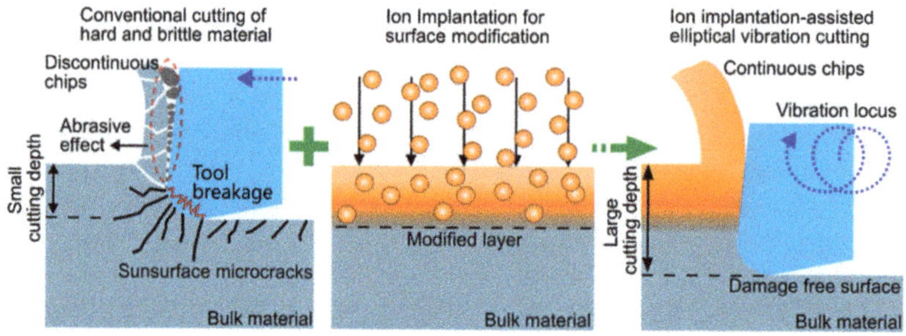

Figure 17. Schematic diagram of multi-field-assisted cutting, which combines ion implantation with tool elliptical vibration [92].

4. Summary

With the rising demand for precision instruments in optics and aerospace, the demand for ultra-precision machining of brittle materials has continued to grow. This paper summarized the brittle materials commonly used in optics, aerospace, and related fields, as well as their deformation and removal mechanisms during ultra-precision cutting. Then, this paper reviewed the development of field-assisted cutting technologies including laser, vibration, magnetic field, and ion implantation, and discussed the material removal mechanisms involved in these processes. The key conclusions are summarized as follows:

(1). This review surveyed the previous research on the machining mechanism of brittle materials during SPDT and synthesized numerical and experimental findings across material classes including single crystals, polycrystalline ceramics and amorphous glass. Because conventional ultra-precision machining faces difficulty in improving machining ability, efficiency, and stability, the review focuses on the effect of laser heating, tool vibration, magnetic field, and ion implantation on broadening the window of ductile-regime machining of brittle materials.

(2). The main field-assisted cutting approaches reviewed are laser-assisted cutting, vibration-assisted cutting, magnetic field-assisted cutting, and ion implantation-assisted cutting. The suitable materials, processing characteristics, advantages and disadvantages of these techniques are summarized in Table 2.

(3). In addition to single field-assisted cutting, this review highlights the emerging direction of multi-field coupling, noting that combined strategies can outperform single-field assistance. The paper discussed the present multi-field coupling-assisted machining strategies and challenges in understanding the mechanisms of multi-field-assisted cutting. The machining mechanism during multi-field-assisted machining is more complex, involving deformation under various thermal, mechanical, and magnetic conditions.

(4). During machining, the selection of the energy field requires comprehensive consideration of material properties, processing costs and applications. The selection of energy fields should be based on the problems encountered in processing, with a comprehensive consideration of the hybrid effects and equipment construction. The summary and review provide ideas and references for research in solving the problems during ultra-precision cutting of brittle materials and exploration of the mechanisms of field-assisted cutting.

https://doi.org/10.3390/mi17030361

Table 2. Comparison between field-assisted cutting technologies.

Assistive Field	Suitable Materials	Advantages	Disadvantages
Laser-assisted cutting	Single-crystal silicon carbide; fused silica; tungsten carbide; single-crystal silicon	Lower cutting forces; improved machinability; higher material removal rates; better surface finish; lower subsurface damage	Introducing thermal expansion and damage; risking tool softening and chemical wear
Vibration-assisted cutting	Stainless steel; calcium fluoride; single-crystal silicon; tungsten alloy	Lower cutting forces; better surface finish; reduced tool wear; higher accuracy in microstructured surface	Introducing complex parameter-dependent surface-generation behavior; reducing machining efficiency
Magnetic field-assisted cutting	Titanium alloy; calcium fluoride	Decreasing the temperature in the deformed region; suppressing the vibration of tool holder; reducing built-up edge and adhesion	Limited suitable materials; magnetization of workpiece material
Ion implantation-assisted cutting	Single-crystal silicon; single-crystal silicon carbide; sapphire	Lower cutting forces; improved surface finish; improved machinability	High cost; low efficiency; potential for introducing undesired elements into the workpiece

5. Future Research Perspectives

(1). Simulation across scales can be effective in the exploration of mechanisms of field-assisted machining. MD simulation can capture phase transitions, amorphization, and dislocation activity in nanometric cutting, but typical domain/time scales limit realistic EVC trajectories and surface formation. Future work can be conducted by coupling MD with the FE method for LAC and EVC, using DEM/phase-field + MD for polycrystalline ceramics where grain boundaries dominate cracking, and validating multiscale models with in-situ measurements during the machining process.

(2). The development of models that predict tool wear, subsurface damage and critical material removal thickness by explicitly coupling tool geometry, workpiece material, and field variables (temperature in LAM, vibration kinematics in EVC, implantation damage gradients). Current indicators such as RTS are useful in predicting material removal mechanisms, but future work should draw more attention to predicting tool wear, subsurface damage and critical material removal thickness in coupling with field variables that reflect thermo-mechanical softening, intermittent contact, and features of modified surface layers.

(3). Multi-field coupling mechanisms (LAC–VAC, implantation–VAC, magnetic–VAC) and synergy/antagonism maps. The review notes that coupled-field effects remain unclear. Future studies should build synergy maps that indicate when a second field enhances or undermines the first. This requires controlled experiments and simulations that isolate coupling terms such as temperature-dependent friction reversal, vibration-modulated heat flux, and substrate interface shielding.

(4). Field-assisted cutting is a promising method to improve the machining performance of various brittle materials. While the distinction between laboratory-scale techniques and industrially mature processes is obvious, laboratory-scale field-assisted cutting techniques often involve custom-built experimental setups, controlled conditions, and small-scale samples, which often face limitations such as complex equipment requirements, difficulty in maintaining process stability, and challenges in integrating with standard CNC systems for reliable operation, repeatable results, and economic viability for large-scale pro-

https://doi.org/10.3390/mi17030361

duction. Therefore, developing industrially mature processes is critical for the application of field-assisted cutting.

Author Contributions: Conceptualization, investigation, and writing—original draft preparation, X.S. and Z.Z.; methodology and software, X.S.; validation, Z.Z.; formal analysis, data curation, supervision and writing—review and editing, C.L. All authors have read and agreed to the published version of the manuscript.

Funding: The work described in this paper was supported by the opening project of the State Key Laboratory of Intelligent Manufacturing Equipment and Technology (Project No. IMETKF2023018).

Institutional Review Board Statement: Not applicable.

Informed Consent Statement: Not applicable.

Data Availability Statement: The original contributions presented in this study are included in the article. Further inquiries can be directed to the corresponding authors.

Conflicts of Interest: Author Xuexiang Sheng is employed by Shandong Harbor Engineering Group Co., Ltd. The remaining authors declare that the research was conducted in the absence of any commercial or financial relationships that could be construed as a potential conflict of interest.

References

1. Chu, J.; Ren, J.; Liu, C.; Zhang, C.; Zhu, B.; Chen, X. The performance of a novel nano-SiO_2 dispersion solutions for the chemical-mechanical polishing of sapphire. *Appl. Surf. Sci.* **2025**, *709*, 163831. [CrossRef]
2. Megson, T.H.G. *Structural and Stress Analysis*; Butterworth-Heinemann: Oxford, UK, 2019.
3. Zhang, X.; Zhang, H.; Liu, C.; Sabet, H.; Xie, S.; Xu, J.; Peng, Q. Development and Evaluation of 0.35-mm-Pitch PET Detectors with Different Reflector Arrangements. *IEEE Trans. Radiat. Plasma Med. Sci.* **2023**, *7*, 802–809. [CrossRef]
4. Wang, R.; Cheung, C.F. 3D Super-resolution Optical Imaging Using Deep Image Prior. In Proceedings of the 2021 International Conference of Optical Imaging and Measurement (ICOIM), Xi'an, China, 27–29 August 2021; pp. 5–8.
5. Wang, Y.; He, Z.; Xie, S.; Wang, R.; Zhang, Z.; Liu, S.; Shang, S.; Zheng, P.; Wang, C. Explainable prediction of surface roughness in multi-jet polishing based on ensemble regression and differential evolution method. *Expert Syst. Appl.* **2024**, *249*, 123578. [CrossRef]
6. Wang, R.; Cheung, C.F.; Liu, C.; Liu, S.; Xiao, H. Self-Supervised Siamese Transformer for Surface Defect Segmentation in Diamond-Wire-Sawn Mono-Crystalline Silicon Wafers. *IEEE Trans. Instrum. Meas.* **2025**, *74*, 1–9. [CrossRef]
7. Wang, Z.; Yang, Y.; Chu, J.; Zang, Y.; She, Z.; Fang, W.; Wang, R. YOLO-LA: Prototype-Based Vision-Language Alignment for Silicon Wafer Defect Pattern Detection. *Micromachines* **2025**, *17*, 67. [CrossRef]
8. Wang, R.; Ji, D.; Liu, C.; Lee, J. V-TimesNet: Vision-Augmented TimesNet for Improved Anomaly Detection in Semiconductor Plasma Dry Etching. Available online: https://ssrn.com/abstract=5955781 (accessed on 23 December 2025).
9. Faggin, F. How we made the microprocessor. *Nat. Electron.* **2018**, *1*, 88. [CrossRef]
10. Auslender, M.; Hava, S. *Single-Crystal Silicon: Electrical and Optical Properties*; Springer: Berlin/Heidelberg, Germany, 2017.
11. Elhebeary, M.; Saif, M.T.A. A novel MEMS stage for in-situ thermomechanical testing of single crystal silicon microbeams under bending. *Extrem. Mech. Lett.* **2018**, *23*, 1–8. [CrossRef]
12. Sell, D.; Yang, J.; Doshay, S.; Zhang, K.; Fan, J.A. Visible Light Metasurfaces Based on Single-Crystal Silicon. *ACS Photonics* **2016**, *3*, 1919–1925. [CrossRef]
13. Wang, R.; Cheung, C.F.; Zang, Y.; Wang, C.; Liu, C. Material removal rate optimization with bayesian optimized differential evolution based on deep learning in robotic polishing. *J. Manuf. Syst.* **2025**, *78*, 178–186. [CrossRef]
14. French, P.; Krijnen, G.; Roozeboom, F. Precision in harsh environments. *Microsyst. Nanoeng.* **2016**, *2*, 16048. [CrossRef]
15. Guo, X.; Xun, Q.; Li, Z.; Du, S. Silicon Carbide Converters and MEMS Devices for High-temperature Power Electronics: A Critical Review. *Micromachines* **2019**, *10*, 406. [CrossRef]
16. Liu, C.; Yip, W.S.; To, S.; Chen, B.; Xu, J. Numerical Investigation on the Effects of Grain Size and Grinding Depth on Nano-Grinding of Cadmium Telluride Using Molecular Dynamics Simulation. *Nanomaterials* **2023**, *13*, 2670. [CrossRef] [PubMed]
17. Liu, C.; Chen, X.; Zhang, J.; Zhang, J.; Chu, J.; Xiao, J.; Xu, J. Molecular dynamic simulation of tool groove wear in nanoscale cutting of silicon. *AIP Adv.* **2020**, *10*, 015327. [CrossRef]
18. Xu, C.; Liang, W.; Ni, C.; Li, X.; Shi, X.; Chu, Z.; Hu, Y.; Yang, Z. Silicone and nano-diamond modified graphene oxide anticorrosive coating. *Surf. Coat. Technol.* **2024**, *479*, 130584. [CrossRef]
19. Zhang, R.; Peng, N.; Ye, H. A perspective on diamond heterojunction devices. *Appl. Phys. Lett.* **2025**, *127*, 070501. [CrossRef]

https://doi.org/10.3390/mi17030361

0. Zhang, X.; Hu, H.; Wang, X.; Luo, X.; Zhang, G.; Zhao, W.; Wang, X.; Liu, Z.; Xiong, L.; Qi, E.; et al. Challenges and strategies in high-accuracy manufacturing of the world's largest SiC aspheric mirror. *Light Sci. Appl.* **2022**, *11*, 310. [CrossRef]

1. Al-Sanabani, F.A.; Al-Qudaimi, N.H. Alumina ceramic for dental applications: A review article. *Am. J. Mater. Res.* **2014**, *1*, 26–34.

2. Jiang, D.; Hulbert, D.; Anselmi-Tamburini, U.; Ng, T.; Land, D.; Muhkerjee, A. Spark plasma sintering and forming of transparent polycrystalline Al2O3 windows and domes. *Proc. SPIE* **2007**, *6545*, 654509.

3. Xiao, W.; Zhang, C.; Gong, X.; Qiu, S.; Wang, J.; Zhang, H.; Luo, W.; Jiang, S.; Li, K.; Zhang, G. Significant Enhancement of Electrocaloric Effect in Ferroelectric Polycrystalline Ceramics Through Grain Boundary Barrier Engineering. *Adv. Funct. Mater.* **2024**, *34*, 2405241. [CrossRef]

4. Tabaru, T.; Shobu, K.; Hirai, H.; Hanada, S. Influences of Al content and secondary phase of Mo5(Si,Al)3 on the oxidation resistance of Al-rich Mo(Si,Al)2-base composites. *Intermetallics* **2003**, *11*, 721–733. [CrossRef]

5. Bourhis, E.L. *Glass Mechanics and Technology*, 2nd ed.; Wiley: Weinheim, Germany, 2014.

6. Chu, J.; Liu, X.; Liu, C.; Zhang, J.; Xiao, J.; Chen, X.; Xu, J. Investigation of dynamic characteristics of fused silica hemispherical resonator with shock and harmonic excitation. *Measurement* **2022**, *204*, 112030. [CrossRef]

7. Sourek, B.; Jirka, V.; Shemelin, V.; Matuska, T. Experimental characterization of glazing with glass prisms. *Sol. Energy* **2017**, *158*, 440–447. [CrossRef]

8. Khegai, A.M.; Alyshev, S.V.; Vakhrushev, A.S.; Riumkin, K.E.; Umnikov, A.A.; Firstov, S.V. Recent advances in Bi-doped silica-based optical fibers: A short review. *J. Non-Cryst. Solids X* **2022**, *16*, 100126. [CrossRef]

9. Chu, J.; Liu, X.; Liu, C.; Zhang, J.; Xiao, J.; Wang, X.; Chen, X.; Xu, J. Fundamental investigation of subsurface damage on the quality factor of hemispherical fused silica shell resonator. *Sens. Actuators A Phys.* **2022**, *335*, 113365. [CrossRef]

0. Yang, D.; Zhang, Z.; Wei, F.; Li, S.; Liu, M.; Lu, Y. Comprehensive Review on Research Status and Progress in Precision Grinding and Machining of BK7 Glasses. *Micromachines* **2024**, *15*, 1021. [CrossRef]

1. Wang, R.; Cheung, C.F.; Wang, C.; Cheng, M.N. Deep learning characterization of surface defects in the selective laser melting process. *Comput. Ind.* **2022**, *140*, 103662. [CrossRef]

2. Wang, R.; Cheung, C.F. Knowledge graph embedding learning system for defect diagnosis in additive manufacturing. *Comput. Ind.* **2023**, *149*, 103912. [CrossRef]

3. Wang, R.; Cheng, M.N.; Loh, Y.M.; Wang, C.; Cheung, C.F. Ensemble learning with a genetic algorithm for surface roughness prediction in multi-jet polishing. *Expert Syst. Appl.* **2022**, *207*, 118024. [CrossRef]

4. Wang, R.; Cheung, C.F. CenterNet-based defect detection for additive manufacturing. *Expert Syst. Appl.* **2022**, *188*, 116000. [CrossRef]

5. Xiao, H.; Yin, S.; Cheung, C.F.; Wang, C. Cracking behavior during scratching brittle materials with different-shaped indenters. *Int. J. Mech. Sci.* **2024**, *268*, 109041. [CrossRef]

6. Wang, R.; Cheung, C.F.; Wang, B.; Zhu, Z.; Tse, D.Y. Enhanced transformer method coupled with transfer learning for surface defect segmentation of myopia control spectacle lenses. *Opt. Express* **2025**, *33*, 13848–13863. [CrossRef] [PubMed]

7. Chen, J.; Liu, C.; Liu, H.; Zhang, B.; To, S. Atomic insight into the speed effect on deformation mechanisms in nano-scratching of monocrystalline iron. *Precis. Eng.* **2024**, *92*, 219–230. [CrossRef]

8. Zhang, H.; Huang, H.; Wang, C.; Zhang, H.; Wu, H. Effects of the indenter orientation and the crystallographic orientation on the scratch characteristics of FeCoCrMnNi high entropy alloy. *Wear 2025*, 564–565, 205694. [CrossRef]

9. Abdulkadir, L.N.; Cheng, Y.; Abou-El-Hossein, K.; Abbas, A.A.S. Process parameter selection for optical silicon considering both experimental and AE results using Taguchi L9 orthogonal design. *Int. J. Adv. Manuf. Technol.* **2019**, *103*, 4355–4367. [CrossRef]

0. Chen, J.; Liu, C.; Yin, T.; Sun, L.; Liu, H.; Zhang, B.; To, S. Surface/subsurface generation and microstructure evolution of iron in ultra-high-speed machining. *Int. J. Mech. Sci.* **2025**, *307*, 110906. [CrossRef]

1. Wang, Y.; Zhang, Y.; Cui, X.; Liang, X.; Li, R.; Wang, R.; Sharma, S.; Liu, M.; Gao, T.; Zhou, Z.; et al. High-speed grinding: From mechanism to machine tool. *Adv. Manuf.* **2024**, *13*, 105–154. [CrossRef]

2. Zang, Y.; Liu, C.; Ke, J.; She, Z.; Chu, J.; Xiao, J.; Zhang, J.; Chen, X.; Xu, J. Mechanisms of strain rate effect on the material removal and surface formation of single-crystal silicon in robotic polishing. *Mater. Sci. Semicond. Process.* **2025**, *199*, 109861. [CrossRef]

3. Mo, H.; Chen, X.; Guo, L.; Zhang, Z.; Chen, X.; Chu, J.; Wang, R. A Transferable Digital Twin-Driven Process Design Framework for High-Performance Multi-Jet Polishing. *Micromachines* **2026**, *17*, 226. [CrossRef]

4. Du, H.; Zhao, D.; Liu, C.; Chen, H.; To, S. Laser-assisted slow tool servo diamond turning of single-crystal silicon for fabricating micro-lens arrays. *J. Manuf. Process.* **2024**, *127*, 174–183. [CrossRef]

5. Ruan, P.; Saxena, D.; Cao, J.; Liu, X.; Wang, R.; Cheung, C.F. NASPrecision: Neural Architecture Search-Driven Multi-Stage Learning for surface roughness prediction in ultra-precision machining. *Expert Syst. Appl.* **2025**, *262*, 125540. [CrossRef]

6. Wang, R.; Cheung, C.F.; Wang, C. Heterogeneous hypergraph learning for analyzing surface defects in additive manufacturing process. *J. Manuf. Syst.* **2024**, *76*, 1–10. [CrossRef]

7. Fu, T.; Liu, S.; Li, P.; Wang, R. Pyramid-based anti-fisheye feature enhancement preprocessing algorithm in torpedo can electrical devices: Application in steel rolling process. *Adv. Manuf.* **2025**. [CrossRef]

https://doi.org/10.3390/mi17030361

48. Puttick, K.; Rudman, M.; Smith, K.; Franks, A.; Lindsey, K. Single-point diamond machining of glasses. *Proc. R. Soc. Lond. A Ma Phys. Sci.* **1989**, *426*, 19–30. [CrossRef]
49. Blake, P.N.; Scattergood, R.O. Ductile-Regime Machining of Germanium and Silicon. *J. Am. Ceram. Soc.* **1990**, *73*, 949–95 [CrossRef]
50. Cai, M.; Li, X.; Rahman, M. Study of the mechanism of nanoscale ductile mode cutting of silicon using molecular dynami simulation. *Int. J. Mach. Tools Manuf.* **2007**, *47*, 75–80. [CrossRef]
51. Liu, H.; Xie, W.; Sun, Y.; Zhu, X.; Wang, M. Investigations on brittle-ductile cutting transition and crack formation in diamor cutting of mono-crystalline silicon. *Int. J. Adv. Manuf. Technol.* **2018**, *95*, 317–326. [CrossRef]
52. Fang, F. Atomic and close-to-atomic scale manufacturing: Perspectives and measures. *Int. J. Extrem. Manuf.* **2020**, *2*, 0302([CrossRef]
53. Ke, J.; Chen, X.; Liu, C.; Xu, G.; She, Z.; Zhang, J.; Xu, J. Ultra-precision cutting characteristics of binderless tungsten carbide b in-heat-process laser-assisted diamond machining. *Int. J. Refract. Met. Hard Mater.* **2023**, *115*, 106311. [CrossRef]
54. Ye, G.; Xue, S.; Jiang, M.; Tong, X.; Dai, L. Modeling periodic adiabatic shear band evolution during high speed machinir Ti-6Al-4V alloy. *Int. J. Plast.* **2013**, *40*, 39–55. [CrossRef]
55. Lu, S.; Li, Z.; Zhang, J.; Zhang, C.; Li, G.; Zhang, H.; Sun, T. Coupled effect of tool geometry and tool-particle position on diamor cutting of SiCp/Al. *J. Mater. Process. Technol.* **2022**, *303*, 117510. [CrossRef]
56. Liu, H.; Xu, X.; Zhang, J.; Liu, Z.; He, Y.; Zhao, W.; Liu, Z. The state of the art for numerical simulations of the effect of th microstructure and its evolution in the metal-cutting processes. *Int. J. Mach. Tools Manuf.* **2022**, *177*, 103890. [CrossRef]
57. He, W.; Liu, C.; Xu, G.; Zhang, J.; Xiao, J.; Chen, X.; Xu, J. Effect of temperature on ductile-to-brittle transition in diamond cuttir of silicon. *Int. J. Adv. Manuf. Technol.* **2021**, *116*, 3447–3462. [CrossRef]
58. Goel, S.; Luo, X.; Agrawal, A.; Reuben, R.L. Diamond machining of silicon: A review of advances in molecular dynami simulation. *Int. J. Mach. Tools Manuf.* **2015**, *88*, 131–164. [CrossRef]
59. Liu, C.; To, S.; Sheng, X.; Wang, R.; Xu, J. Atomic simulation of crystal orientation and workpiece composition effect o nano-scratching of SiGe alloy. *Discover Nano* **2023**, *18*, 91. [CrossRef]
60. Liu, C.; Chu, J.; Ke, J.; Chen, X.; Zhang, J.; Xiao, J.; Xu, J. Molecular Dynamic Simulation of Micro-Structured Diamond Tool i Silicon Carbide Cutting. *J. Micro Nano Manuf.* **2021**, *9*, 021002. [CrossRef]
61. Liu, C.; Zhuang, Z.; Chen, J.; Yip, W.S.; To, S. Atomic-scale investigation of Pt composition on deformation mechanism of Au alloy during nano-scratching process. *Surf. Interfaces* **2023**, *40*, 103126. [CrossRef]
62. Liu, C.; Zhang, J.; Zhang, J.; Chen, X.; He, W.; Xiao, J.; Xu, J. Influence of micro grooves of diamond tool on silicon cutting: molecular dynamic study. *Mol. Simul.* **2020**, *46*, 92–101. [CrossRef]
63. Meng, B.; Qiu, P.; Yuan, D.; Xu, S. Influence of microstructure on the diamond-machinability of hot-pressed silicon carbide: molecular dynamics study. *Ceram. Int.* **2019**, *45*, 22872–22879. [CrossRef]
64. Liu, C.; Chu, J.; Chen, X.; Xiao, J.; Xu, J. Molecular dynamics simulation on structure evolution of silica glass in nano-cutting a high temperature. *Mol. Simul.* **2020**, *46*, 957–965. [CrossRef]
65. Fang, F.; Wu, H.; Liu, Y. Modelling and experimental investigation on nanometric cutting of monocrystalline silicon. *Int. J. Mac Tools Manuf.* **2005**, *45*, 1681–1686. [CrossRef]
66. Zhang, L.; Zhao, H.; Ma, Z.; Huang, H.; Shi, C.; Zhang, W. A study on phase transformation of monocrystalline silicon due t ultra-precision polishing by molecular dynamics simulation. *AIP Adv.* **2012**, *2*, 042116. [CrossRef]
67. Fang, F.; Wu, H.; Zhou, W.; Hu, X. A study on mechanism of nano-cutting single crystal silicon. *J. Mater. Process. Technol.* **200** *184*, 407–410. [CrossRef]
68. Chavoshi, S.Z.; Goel, S.; Luo, X. Molecular dynamics simulation investigation on the plastic flow behaviour of silicon durin nanometric cutting. *Model. Simul. Mater. Sci. Eng.* **2016**, *24*, 015002. [CrossRef]
69. Zhao, H.; Shi, C.; Zhang, P.; Zhang, L.; Huang, H.; Yan, J. Research on the effects of machining-induced subsurface damages o mono-crystalline silicon via molecular dynamics simulation. *Appl. Surf. Sci.* **2012**, *259*, 66–71. [CrossRef]
70. Mylvaganam, K.; Zhang, L.C. Nanotwinning in monocrystalline silicon upon nanoscratching. *Scr. Mater.* **2011**, *65*, 214–21 [CrossRef]
71. Liu, B.; Xu, Z.; Chen, C.; Pang, K.; Wang, Y.; Ruan, Q. Effect of tool edge radius on material removal mechanism of single-cryst silicon: Numerical and experimental study. *Comput. Mater. Sci.* **2019**, *163*, 127–133. [CrossRef]
72. Ameli Kalkhoran, S.N.; Vahdati, M.; Yan, J. Molecular Dynamics Investigation of Nanometric Cutting of Single-Crystal Silico Using a Blunt Tool. *JOM* **2019**, *71*, 4296–4304. [CrossRef]
73. Wu, Z.; Zhang, L.; Yang, S.; Wu, C. Effects of grain size and protrusion height on the surface integrity generation in th nanogrinding of 6H-SiC. *Tribol. Int.* **2022**, *171*, 107563. [CrossRef]
74. Liu, C.; Chu, J.; Zhang, J.; Zhang, J.; Chen, X.; Xiao, J.; Xu, J. Effect of tool rake angle on the material removal mechanism transitio of single-crystal silicon: A molecular dynamics study. *Int. J. Adv. Manuf. Technol.* **2021**, *115*, 3631–3644. [CrossRef]

https://doi.org/10.3390/mi1703036

75. Liu, C.; Zhu, Z.; Chen, J.; Wang, R. Effect of grain size on nano-scratching of RB-SiC via molecular dynamics simulation. *Vacuum* **2026**, *245*, 115001. [CrossRef]
76. She, Z.; Liu, C.; Ke, J.; Zang, Y.; Zhang, J.; Chen, X.; Xu, J. Material removal mechanism of cryogenic-laser assisted cutting for SiCp/Al composites. *Opt. Laser Technol.* **2025**, *192*, 113396. [CrossRef]
77. Tan, Y.; Yang, D.; Sheng, Y. Discrete element method (DEM) modeling of fracture and damage in the machining process of polycrystalline SiC. *J. Eur. Ceram. Soc.* **2009**, *29*, 1029–1037. [CrossRef]
78. Liu, Y.; Li, B.; Kong, L. A molecular dynamics investigation into nanoscale scratching mechanism of polycrystalline silicon carbide. *Comput. Mater. Sci.* **2018**, *148*, 76–86. [CrossRef]
79. Goel, S.; Kovalchenko, A.; Stukowski, A.; Cross, G. Influence of microstructure on the cutting behaviour of silicon. *Acta Mater.* **2016**, *105*, 464–478. [CrossRef]
80. Zhao, L.; Hu, W.; Zhang, Q.; Zhang, J.; Zhang, J.; Sun, T. Atomistic origin of brittle-to-ductile transition behavior of polycrystalline 3C–SiC in diamond cutting. *Ceram. Int.* **2021**, *47*, 23895–23904. [CrossRef]
81. Yan, J.; Zhang, Z.; Kuriyagawa, T. Mechanism for material removal in diamond turning of reaction-bonded silicon carbide. *Int. J. Mach. Tools Manuf.* **2009**, *49*, 366–374. [CrossRef]
82. Guo, W.; Yu, Q.; Wang, G.; Fu, S.; Liu, C.; Chen, X. Effect of Grain Size on Nanometric Cutting of Polycrystalline Silicon via Molecular Dynamics Simulation. *Micromachines* **2024**, *15*, 767. [CrossRef]
83. Guo, X.; Zhai, C.; Kang, R.; Jin, Z. The mechanical properties of the scratched surface for silica glass by molecular dynamics simulation. *J. Non-Cryst. Solids* **2015**, *420*, 1–6. [CrossRef]
84. Komanduri, R. On Material Removal Mechanisms in Finishing of Advanced Ceramics and Glasses. *CIRP Ann.* **1996**, *45*, 509–514. [CrossRef]
85. Hermansen, C.; Matsuoka, J.; Yoshida, S.; Yamazaki, H.; Kato, Y.; Yue, Y.Z. Densification and plastic deformation under microindentation in silicate glasses and the relation to hardness and crack resistance. *J. Non-Cryst. Solids* **2013**, *364*, 40–43. [CrossRef]
86. Wang, W.; Yao, P.; Wang, J.; Huang, C.; Zhu, H.; Zou, B.; Liu, H.; Yan, J. Crack-free ductile mode grinding of fused silica under controllable dry grinding conditions. *Int. J. Mach. Tools Manuf.* **2016**, *109*, 126–136. [CrossRef]
87. Liu, H.; Shi, Y.; Youngman, R.E.; Huang, L. Role of densification in deformation behaviors of model metallic glasses under 3-D nanoindentation studied in molecular dynamics simulation. *J. Non-Cryst. Solids* **2024**, *638*, 123071. [CrossRef]
88. Roostaei, H.; Movahhedy, M.R. Analysis of Heat Transfer in Laser Assisted Machining of Slip Cast Fused Silica Ceramics. *Procedia CIRP* **2016**, *46*, 571–574. [CrossRef]
89. Yuan, F.; Huang, L. Brittle to ductile transition in densified silica glass. *Sci. Rep.* **2014**, *4*, 5035. [CrossRef] [PubMed]
90. Liu, C.; To, S.; Sheng, X.; Xu, J. Molecular dynamics simulation on crystal defects of single-crystal silicon during elliptical vibration cutting. *Int. J. Mech. Sci.* **2023**, *244*, 108072. [CrossRef]
91. Liu, C.; Chen, X.; Ke, J.; She, Z.; Zhang, J.; Xiao, J.; Xu, J. Numerical investigation on subsurface damage in nanometric cutting of single-crystal silicon at elevated temperatures. *J. Manuf. Process.* **2021**, *68*, 1060–1071. [CrossRef]
92. Ke, J.; Zhang, J.; Chen, X.; Liu, C.; Long, G.; Sun, H.; Xu, J. Investigation on the material removal mechanism in ion implantation-assisted elliptical vibration cutting of hard and brittle material. *Int. J. Mach. Tools Manuf.* **2024**, *203*, 104220. [CrossRef]
93. Wang, R.; Cheung, C.F.; Wang, C. Unsupervised Defect Segmentation in Selective Laser Melting. *IEEE Trans. Instrum. Meas.* **2023**, *72*, 2520010. [CrossRef]
94. Dumitrescu, P.; Koshy, P.; Stenekes, J.; Elbestawi, M.A. High-power diode laser assisted hard turning of AISI D2 tool steel. *Int. J. Mach. Tools Manuf.* **2006**, *46*, 2009–2016. [CrossRef]
95. García Navas, V.; Arriola, I.; Gonzalo, O.; Leunda, J. Mechanisms involved in the improvement of Inconel 718 machinability by laser assisted machining (LAM). *Int. J. Mach. Tools Manuf.* **2013**, *74*, 19–28. [CrossRef]
96. Anderson, M.; Patwa, R.; Shin, Y.C. Laser-assisted machining of Inconel 718 with an economic analysis. *Int. J. Mach. Tools Manuf.* **2006**, *46*, 1879–1891. [CrossRef]
97. Leshock, C.E.; Kim, J.N.; Shin, Y.C. Plasma enhanced machining of Inconel 718: Modeling of workpiece temperature with plasma heating and experimental results. *Int. J. Mach. Tools Manuf.* **2001**, *41*, 877–897. [CrossRef]
98. Song, H.; Li, J.; Dan, J.; Ren, G.; Xiao, J.; Xu, J. Experimental analysis and evaluation of the cutting performance of tools in laser-assisted machining of fused silica. *Precis. Eng.* **2019**, *56*, 191–202. [CrossRef]
99. Maity, K.P.; Swain, P.K. An experimental investigation of hot-machining to predict tool life. *J. Mater. Process. Technol.* **2008**, *198*, 344–349. [CrossRef]
100. Amin, A.K.M.N.; Dolah, S.B.; Mahmud, M.B.; Lajis, M.A. Effects of workpiece preheating on surface roughness, chatter and tool performance during end milling of hardened steel D2. *J. Mater. Process. Technol.* **2008**, *201*, 466–470. [CrossRef]

https://doi.org/10.3390/mi17030361

101. Ravindra, D.; Ghantasala, M.K.; Patten, J. Ductile mode material removal and high-pressure phase transformation in silico during micro-laser assisted machining. *Precis. Eng.* **2012**, *36*, 364–367. [CrossRef]
102. Ke, J.; Chen, X.; Liu, C.; Zhang, J.; Yang, H.; Xu, J. Enhancing the ductile machinability of single-crystal silicon by laser-assiste diamond cutting. *Int. J. Adv. Manuf. Technol.* **2022**, *118*, 3265–3282. [CrossRef]
103. Ke, J.; Fu, Y.; Liu, C.; Zhang, J.; Chen, X.; Xu, J. Investigation on system design methodology and cutting force optimization i laser-assisted diamond machining of single-crystal silicon. *J. Manuf. Process.* **2024**, *115*, 1–17. [CrossRef]
104. Lin, C.; Chen, X.; He, W.; Xu, G.; Liu, C.; Zhang, J.; Xu, J. Experimental investigation on the ductile machinability of fused silic during in-situ laser assisted diamond cutting. *J. Manuf. Process.* **2022**, *84*, 383–393. [CrossRef]
105. Zhang, J.; Fu, Y.; Chen, X.; Shen, Z.; Zhang, J.; Xiao, J.; Xu, J. Investigation of the material removal process in in-situ laser-assiste diamond cutting of reaction-bonded silicon carbide. *J. Eur. Ceram. Soc.* **2023**, *43*, 2354–2365. [CrossRef]
106. Langan, S.M.; Ravindra, D.; Mann, A.B. Mitigation of damage during surface finishing of sapphire using laser-assisted machinin *Precis. Eng.* **2019**, *56*, 1–7. [CrossRef]
107. Kim, J.D.; Lee, S.J.; Suh, J. Characteristics of laser assisted machining for silicon nitride ceramic according to machining parameter *J. Mech. Sci. Technol.* **2011**, *25*, 995–1001. [CrossRef]
108. Chen, X.; Liu, C.; Ke, J.; Zhang, J.; Shu, X.; Xu, J. Subsurface damage and phase transformation in laser-assisted nanometri cutting of single crystal silicon. *Mater. Des.* **2020**, *190*, 108524. [CrossRef]
109. Zou, C.; Shi, G.; Meng, S.; Kong, D.; Yao, D. Experimental study on in-situ laser-assisted diamond turning of single crysta germanium. *Precis. Eng.* **2025**, *94*, 608–622. [CrossRef]
110. Ke, J.; Liu, C.; Wang, C.; Yu, X.; Hu, Y.; Zhang, J.; Chen, X.; Xu, J. Analytical force modeling for laser-assisted diamond machinin of brittle materials. *Int. J. Mech. Sci.* **2025**, *301*, 110494. [CrossRef]
111. Lin, C.; Zhang, H.; Zhou, Y.; Wei, J.; Zhang, J.; Chen, X.; Xu, J. The mechanism and machinability of glass-ceramics durin in-process laser assisted cutting. *Ceram. Int.* **2025**, *51*, 62137–62147. [CrossRef]
112. Xiong, X.; Yang, M.; Liu, C.; Li, X.; Tang, D. Thermal conductivity of cross-linked polyethylene from molecular dynamic simulation. *J. Appl. Phys.* **2017**, *122*, 035104. [CrossRef]
113. Chen, B.; Wu, Z.; Liu, C.; Zhang, J.; Chen, X.; Xiao, J.; Xu, J. Molecular dynamics simulation for nanometric cutting of NiTi shap memory alloys at elevated temperatures. *J. Manuf. Process.* **2024**, *124*, 581–589. [CrossRef]
114. Mohammadi, H.; Poyraz, H.B.; Ravindra, D.; Patten, J.A. Surface finish improvement of an unpolished silicon wafer usin micro-laser assisted machining. *Int. J. Abras. Technol.* **2015**, *7*, 107–121. [CrossRef]
115. Dai, H.; Li, S.; Chen, G. Comparison of subsurface damages on mono-crystalline silicon between traditional nanoscale machining and laser-assisted nanoscale machining via molecular dynamics simulation. *Nucl. Instrum. Methods Phys. Res. Sect. B* **2018**, *414* 61–67. [CrossRef]
116. You, K.; Fang, F.; Yan, G. Surface generation of tungsten carbide in laser-assisted diamond turning. *Int. J. Mach. Tools Manuf.* **2021** *168*, 103770. [CrossRef]
117. Ye, G.; Zhang, P.; Zhang, J.; Zhang, Y.; Huang, X. Study on nanometer cutting mechanism of single crystal silicon at differen temperatures. *J. Manuf. Process.* **2023**, *93*, 275–286. [CrossRef]
118. Chavoshi, S.Z.; Luo, X. Atomic-scale characterization of occurring phenomena during hot nanometric cutting of single crysta 3C–SiC. *RSC Adv.* **2016**, *6*, 71409. [CrossRef]
119. Meng, B.; Yuan, D.; Xu, S. Study on strain rate and heat effect on the removal mechanism of SiC during nano-scratching process by molecular dynamics simulation. *Int. J. Mech. Sci.* **2019**, *151*, 724–732. [CrossRef]
120. Metcalf, T.H.; Liu, X.; Jernigan, G.; Culbertson, J.C.; Abernathy, M.; Molina-Ruiz, M.; Hellman, F. Internal friction measurements of low energy excitations in amorphous germanium thin films. *J. Alloys Compd.* **2021**, *856*, 157616. [CrossRef]
121. Liu, C.; Xu, W.; Zhang, J.; Xiao, J.; Chen, X.; Xu, J. Numerical investigation on the temperature effect in nanometric cutting of polycrystalline silicon. *Int. J. Mech. Sci.* **2022**, *220*, 107172. [CrossRef]
122. Liu, C.; Yip, W.S.; Chen, J.; Wang, R.; Xu, J.; To, S. Atomic simulation of the temperature effect on fabrication mechanism of micro-structured surface on single-crystal silicon. *J. Manuf. Process.* **2025**, *133*, 238–248. [CrossRef]
123. Langana, S.M.; Ravindra, D.; Mann, A.B. Process parameter effects on residual stress and phase purity after microlaser-assisted machining of silicon. *Mater. Manuf. Process.* **2018**, *33*, 1578–1586. [CrossRef]
124. Chen, X.; Chu, J.; Zhu, Z.; Liu, C. Effect of temperature on surface morphology of single-crystal silicon in nanometric cutting *Appl. Surf. Sci.* **2025**, *684*, 161957. [CrossRef]
125. Liu, C.; Ke, J.; Yin, T.; Yip, W.S.; Zhang, J.; To, S.; Xu, J. Cutting mechanism of reaction-bonded silicon carbide in laser-assisted ultra-precision machining. *Int. J. Mach. Tools Manuf.* **2024**, *203*, 104219. [CrossRef]
126. Liu, Z.; Lin, B.; Liang, X.; Du, A. Study on the effect of laser-assisted machining on tool wear based on molecular dynamics simulation. *Diam. Relat. Mater.* **2020**, *109*, 108022. [CrossRef]
127. Niu, Y.; Zhao, D.; Wang, S.; Li, S.; Wang, Z.; Zhao, H. Investigations on thermal effects on scratch behavior of monocrystalline silicon via molecular dynamics simulation. *Mater. Today Commun.* **2021**, *26*, 102042. [CrossRef]

28. Skelton, R.C. Turning with an oscillating tool. *Int. J. Mach. Tool Des. Res.* **1968**, *8*, 239–259. [CrossRef]
29. Zhou, M.; Eow, Y.T.; Ngoi, B.K.A.; Lim, E.N. Vibration-Assisted Precision Machining of Steel with PCD Tools. *Mater. Manuf. Process.* **2003**, *18*, 825–834. [CrossRef]
30. Xiao, M.; Karube, S.; Sato, K. Analysis of chatter suppression in vibration cutting. *Int. J. Mach. Tools Manuf.* **2002**, *42*, 1677–1685. [CrossRef]
31. Zhou, M.; Ngoi, B.K.A.; Yusoff, M.N.; Wang, X.J. Tool wear and surface finish in diamond cutting of optical glass. *J. Mater. Process. Technol.* **2006**, *174*, 29–33. [CrossRef]
32. Moriwaki, T.; Shamoto, E. Ultrasonic Elliptical Vibration Cutting. *CIRP Ann.* **1995**, *44*, 31–34. [CrossRef]
33. Wang, J.; Guo, P. Effects of vibration trajectory on ductile-to-brittle transition in vibration cutting of single crystal silicon using a non-resonant tool. In Proceedings of the 4th CIRP Conference on Surface Integrity (CSI 2018), Tianjin, China, 11–13 July 2018; pp. 289–292.
34. Zhang, J.; Han, L.; Zhang, J.; Liu, H.; Yan, Y.; Sun, T. Brittle-to-ductile transition in elliptical vibration-assisted diamond cutting of reaction-bonded silicon carbide. *J. Manuf. Process.* **2019**, *45*, 670–681. [CrossRef]
35. Nath, C.; Rahman, M.; Neo, K.S. A study on the effect of tool nose radius in ultrasonic elliptical vibration cutting of tungsten carbide. *J. Mater. Process. Technol.* **2009**, *209*, 5830–5836. [CrossRef]
36. Zhang, J.; Zhang, J.; Liu, C.; Chen, X.; Xiao, J.; Xu, J. Machinability of single crystal calcium fluoride by applying elliptical vibration diamond cutting. *Precis. Eng.* **2020**, *66*, 306–314. [CrossRef]
37. Wang, T.; Liang, X.; Wang, B.; Liu, K.; Li, T.; Cai, Y.; Liu, Z. Ultrasonic vibration-assisted machining in aerospace composite materials: Principle, technology, challenges. *Compos. Part A Appl. Sci. Manuf.* **2025**, *199*, 109250. [CrossRef]
38. Zhang, Y.; Hu, Z.; Yu, Y.; Xu, X.; Zeng, W.; Zhong, W.; Blunt, L.; Jiang, X. Crystal-orientation-dependent nanoscale machining mechanisms in ultrasonic vibration-assisted scratching sapphire. *Appl. Surf. Sci.* **2026**, *719*, 165069. [CrossRef]
39. Xu, W.; Zhang, L. Heat effect on the material removal in the machining of fibre-reinforced polymer composites. *Int. J. Mach. Tools Manuf.* **2019**, *140*, 1–11. [CrossRef]
40. Zhao, L.; Zhang, J.; Zhang, J.; Hartmaier, A. Atomistic investigation of machinability of monocrystalline 3C–SiC in elliptical vibration-assisted diamond cutting. *Ceram. Int.* **2021**, *47*, 2358–2366. [CrossRef]
41. Zhu, B.; Zhao, D.; Zhao, H.; Guan, J.; Hou, P.; Wang, S.; Qian, L. A study on the surface quality and brittle–ductile transition during the elliptical vibration-assisted nanocutting process on monocrystalline silicon via molecular dynamic simulations. *RSC Adv.* **2017**, *7*, 4179–4189. [CrossRef]
42. Chen, Y.; Hu, Z.; Jin, J.; Li, L.; Yu, Y.; Peng, Q.; Xu, X. Molecular dynamics simulations of scratching characteristics in vibration-assisted nano-scratch of single-crystal silicon. *Appl. Surf. Sci.* **2021**, *551*, 149451. [CrossRef]
43. Wu, H.; Zhang, W.; Chen, Z.; Zeng, Q.; Sun, L.; Zeng, J.; Zhang, S.; Liu, J.; Xiao, C.; Wu, Y. Mechanisms and effects of in-plane ultrasonic-assisted nanoscratching on single-crystal SiC. *Appl. Surf. Sci.* **2026**, *720*, 165222. [CrossRef]
44. Zhang, J.; Zhang, J.; Cui, T.; Hao, Z.; Zahrani, A.A. Sculpturing of single crystal silicon microstructures by elliptical vibration cutting. *J. Manuf. Process.* **2017**, *29*, 389–398. [CrossRef]
45. Nath, C.; Rahman, M.; Neo, K.S. Modeling of the Effect of Machining Parameters on Maximum Thickness of Cut in Ultrasonic Elliptical Vibration Cutting. *J. Manuf. Sci. Eng.* **2011**, *133*, 011007. [CrossRef]
46. Zhang, X.; Kumar, A.S.; Rahman, M.; Nath, C.; Liu, K. Experimental study on ultrasonic elliptical vibration cutting of hardened steel using PCD tools. *J. Mater. Process. Technol.* **2011**, *211*, 1701–1709. [CrossRef]
47. Zhang, X. A Study of Elliptical Vibration Cutting in Ultra Precision Machining. Ph.D. Thesis, National University of Singapore, Singapore, 2012.
48. Nath, C.; Rahman, M. Effect of machining parameters in ultrasonic vibration cutting. *Int. J. Mach. Tools Manuf.* **2008**, *48*, 965–974. [CrossRef]
49. Huang, W.; Yu, D.; Zhang, X.; Zhang, M.; Chen, D. Ductile-regime machining model for ultrasonic elliptical vibration cutting of brittle materials. *J. Manuf. Process.* **2018**, *36*, 68–76. [CrossRef]
50. Dai, H.; Du, H.; Chen, J.; Chen, G. Influence of elliptical vibration on the behavior of silicon during nanocutting. *Int. J. Adv. Manuf. Technol.* **2019**, *102*, 3597–3612. [CrossRef]
51. Liu, C.; Zhang, J.; Zhang, J.; Chen, X.; Xiao, J.; Xu, J. A simulation investigation on elliptical vibration cutting of single-crystal silicon. *Int. J. Adv. Manuf. Technol.* **2020**, *108*, 2231–2243. [CrossRef]
52. Zhang, D.; Li, H.; Tang, J.; Huang, W. Cutting mechanism and surface characteristics of during ultrasonic vibration-assisted cutting iron alloy: An atomistic investigation. *Tribol. Int.* **2026**, *214*, 111275. [CrossRef]
53. Li, C.; Yang, R.; Gao, C.; Liu, G.; Zhang, Y.; Zakharov, O.; Wang, N.; Geng, Y. Atomic-scale insight into damage and removal behaviors during ultrasonic elliptical vibration-assisted grinding of CaF2 crystals. *J. Mater. Res. Technol.* **2025**, *36*, 1349–1361. [CrossRef]

https://doi.org/10.3390/mi17030361

154. Liu, C.; Zhang, J.; Zhang, J.; Chu, J.; Chen, X.; Xiao, J.; Xu, J. Numerical investigation on material removal mechanism in ellipti vibration cutting of single-crystal silicon. *Mater. Sci. Semicond. Process.* **2021**, *134*, 106019. [CrossRef]
155. Yip, W.S.; To, S. Reduction of material swelling and recovery of titanium alloys in diamond cutting by magnetic field assistan *J. Alloys Compd.* **2017**, *722*, 525–531. [CrossRef]
156. Yip, W.S.; He, T.; Wang, H.; To, S. Facile fabrication of hierarchical micro/nano aluminum alloy surfaces with enhanc hydrophobicity and surface quality by single point diamond cutting under a magnetic field influence. *J. Mater. Res. Technol.* **202** *30*, 7946–7957. [CrossRef]
157. Guo, Y.; Zhan, J.; Lee, Y.J.; Lu, W.F.; Wang, H. Predictive modelling for enhanced scratching of brittle ceramics with magne plasticity. *Int. J. Mech. Sci.* **2023**, *249*, 108272. [CrossRef]
158. Guo, Y.; Lee, Y.J.; Zhang, Y.; Sorkin, A.; Manzhos, S.; Wang, H. Effect of a weak magnetic field on ductile–brittle transition micro-cutting of single-crystal calcium fluoride. *J. Mater. Sci. Technol.* **2022**, *112*, 96–113. [CrossRef]
159. Guo, Y.; Zhan, J.; Lu, W.F.; Wang, H. Mechanism in scratching of calcium fluoride with magneto-plasticity. *Int. J. Mech. Sci.* **202** *263*, 108768. [CrossRef]
160. Li, D.; Yip, W.S.; Cao, H.; Zhang, H.; Tang, Y.M.; To, S. Chatter suppression in diamond turning using magnetic field assistan *J. Mater. Process. Technol.* **2023**, *321*, 118150. [CrossRef]
161. Jiang, F.; Yan, L.; Huang, Y.; Xu, X. Review on Magnetic Field Assisted Machining Technology. *J. Mech. Eng.* **2016**, *52*, 1– [CrossRef]
162. Pelaz, L.; Marqués, L.A.; Barbolla, J. Ion-beam-induced amorphization and recrystallization in silicon. *J. Appl. Phys.* **2004**, *9* 5947–5976. [CrossRef]
163. Li, W.; Zhan, X.; Song, X.; Si, S.; Chen, R.; Liu, J.; Wang, Z.; He, J.; Xiao, X. A Review of Recent Applications of Ion Bea Techniques on Nanomaterial Surface Modification: Design of Nanostructures and Energy Harvesting. *Small* **2019**, *15*, 190182 [CrossRef] [PubMed]
164. Xu, Z.; Fang, F.; Fu, Y.; Zhang, S.; Han, T.; Li, J. Fabrication of micro/nano-structures using focused ion beam implantation an XeF2gas-assisted etching. *J. Micromech. Microeng.* **2009**, *19*, 054003. [CrossRef]
165. Fang, F.; Chen, Y.; Zhang, X.; Hu, X.; Zhang, G. Nanometric cutting of single crystal silicon surfaces modified by ion implantatio *CIRP Ann.* **2011**, *60*, 527–530. [CrossRef]
166. Chen, Y.; Fang, F.; Zhang, X.; Hu, X. Molecular dynamics of nanometric processing of ion implanted monocrystalline silico surfaces. *Trans. Tianjin Univ.* **2014**, *20*, 203–209. [CrossRef]
167. Wang, J.; Chen, R.; Zhang, X.; Fang, F. Study on machinability of silicon irradiated by swift ions. *Precis. Eng.* **2018**, *51*, 577–58 [CrossRef]
168. To, S.; Wang, H.; Jelenković, E.V. Enhancement of the machinability of silicon by hydrogen ion implantation for ultra-precisio micro-cutting. *Int. J. Mach. Tools Manuf.* **2013**, *74*, 50–55. [CrossRef]
169. Fan, Y.; Xu, Z.; Song, Y.; Dong, B.; Xue, Z.; Liu, B.; Liu, L.; Tian, D. Nano material removal mechanism of 4H-SiC in io implantation-assisted machining. *Comput. Mater. Sci.* **2021**, *200*, 110837. [CrossRef]
170. Liu, B.; Xu, Z.; Wang, Y.; Gao, X.; Kong, R. Effect of ion implantation on material removal mechanism of 6H-SiC in nano-cuttin A molecular dynamics study. *Comput. Mater. Sci.* **2020**, *174*, 109476. [CrossRef]
171. Tanaka, H.; Shimada, S. Damage-free machining of monocrystalline silicon carbide. *CIRP Ann.* **2013**, *62*, 55–58. [CrossRef]
172. Xu, Z.; Liu, L.; He, Z.; Tian, D.; Hartmaier, A.; Zhang, J.; Luo, X.; Rommel, M.; Nordlund, K.; Zhang, G.; et al. Nanocuttin mechanism of 6H-SiC investigated by scanning electron microscope online observation and stress-assisted and ion implan assisted approaches. *Int. J. Adv. Manuf. Technol.* **2020**, *106*, 3869–3880. [CrossRef]
173. Fan, Y.; Song, Y.; Xu, Z.; Dong, B.; Wu, J.; Rommel, M.; Zhang, K.; Zhao, J.; Zhu, R.; Li, B.; et al. Molecular dynamics simulation c color centers in silicon carbide by helium and dual ion implantation and subsequent annealing. *Ceram. Int.* **2021**, *47*, 24534–2454 [CrossRef]
174. Wang, J.; Zhang, X.; Fang, F.; Chen, R. Diamond cutting of micro-structure array on brittle material assisted by multi-io implantation. *Int. J. Mach. Tools Manuf.* **2019**, *137*, 58–66. [CrossRef]
175. Tan, Y.; Yip, W.S.; Zhao, T.; To, S.; Zhao, Z. Subsurface damage and brittle fracture suppression of monocrystalline germaniu in ultra-precision machining by multiple ion implantation surface modification. *J. Mater. Process. Technol.* **2024**, *334*, 11864 [CrossRef]
176. Xing, Y.; Liu, Y.; Liu, C.; Li, Y.; Yin, T.; Yin, S.; Sun, Z.; Zhu, Z.; Xue, C.; Yip, W.S.; et al. An innovative multi-energy field-assiste ultra-precision machining technology: In-situ laser-magnetic dual-field assisted diamond cutting. *Int. J. Extrem. Manuf.* **202** *8*, 025103. [CrossRef]
177. Wang, J.; Fang, F.; Yan, G.; Guo, Y. Study on Diamond Cutting of Ion Implanted Tungsten Carbide with and Without Ultrasoni Vibration. *Nanomanuf. Metrol.* **2019**, *2*, 177–185. [CrossRef]

https://doi.org/10.3390/mi1703036

78. Zhang, J.; Li, X.; Ma, S.; Lu, Y.; Tian, H.; Xiao, J.; Xu, J. High efficiency fabrication of Si microlenses by applying in-situ laser and ultrasonic vibration hybrid diamond cutting. *Precis. Eng.* **2025**, *93*, 324–333. [CrossRef]
79. Liu, C.; He, W.; Chu, J.; Zhang, J.; Chen, X.; Xiao, J.; Xu, J. Molecular Dynamics Simulation on Cutting Mechanism in the Hybrid Machining Process of Single-Crystal Silicon. *Nanoscale Res. Lett.* **2021**, *16*, 66. [CrossRef] [PubMed]

https://doi.org/10.3390/mi17030361

MDPI AG
Grosspeteranlage 5
4052 Basel
Switzerland
Tel.: +41 61 683 77 34

Micromachines Editorial Office
E-mail: micromachines@mdpi.com
www.mdpi.com/journal/micromachines

www.ingramcontent.com/pod-product-compliance
Lightning Source LLC
Chambersburg PA
CBHW062035220926
51850CB00010B/2563
* 9 7 8 3 7 2 5 8 7 6 8 1 5 *